THE HUMAN SUCCESS PLAYBOOK

A LEADERSHIP TALE FOR THE AGE OF AI

ROBIN DANIELS   RASMUS HOLST   NINA CARØE

WITH
CHRIS WESTFALL

PRAISE FOR THE HUMAN SUCCESS PLAYBOOK

———

"Most books about AI talk about technology. This one talks about *leverage*. *The Human Success Playbook* forces leaders to confront the only question that matters right now: are you scaling software, or are you scaling humans? If you run a company and care about performance, this isn't optional reading."

"I see leaders every day who are **lost in the noise of digital transformation**. This book is the peace offering they need—a way to reconcile the demands of the bottom line with the needs of the human beings who make that bottom line possible."

"Most companies treat learning, engagement, and performance as separate silos. *The Human Success Playbook* smashes those walls with The Trifecta. This isn't HR theory—it's **a battle plan for CEOs** who want to unlock the hidden potential of their workforce to out-compete AI-native rivals."

- LOU DIAMOND, AUTHOR OF *SPEAK EASY* AND HOST, THRIVE LOUD! PODCAST

"This book captures **the global tension in the age of AI** and what leadership must consider. How can we bridge the gap between Scandinavian *tryghed* (security) and Silicon Valley's unbridled growth -- while meeting humanity in this moment and shaping our collective future? Read this book and find out."

- JENNIFER TACHEFF, FOUNDER OF MANIFEST ADVISORS, INNOVATION STRATEGIST AND SXSW PITCH COACH

"An insider's view into how to **build a better culture** by understanding the human aspect at the center of every organization. This *Playbook* asks how we avoid scaling a broken system and instead build one that will sustain the future—one that understands the power of the ecosystem where HR means human relationships, and AI is collaborating towards hybrid intelligence.

The Human Success Playbook shows leaders how to leverage our human advantage in an AI-driven world rather than compete against it. It provides **the blueprint for building a collaborative infrastructure** that creates a better future for everyone—not just board members. A refreshing take on a corporate tale."

- JEANETTE BRONÉE, CULTURE STRATEGIST, TEDX SPEAKER AND AUTHOR OF *THE SELF-CARE MINDSET®*

The HUMAN SUCCESS PLAYBOOK:
A Leadership Tale for the Age of AI

ISBN: 979-8-9954734-0-4 Print
979-8-9954734-1-1 eBook
Cover design by Mr. Josh

Interior Design and Imagery: Chris Westfall, Zensai
Some images were developed and/or modified using Google Gemini
Printed Globally via IngramSpark

TABLE OF CONTENTS

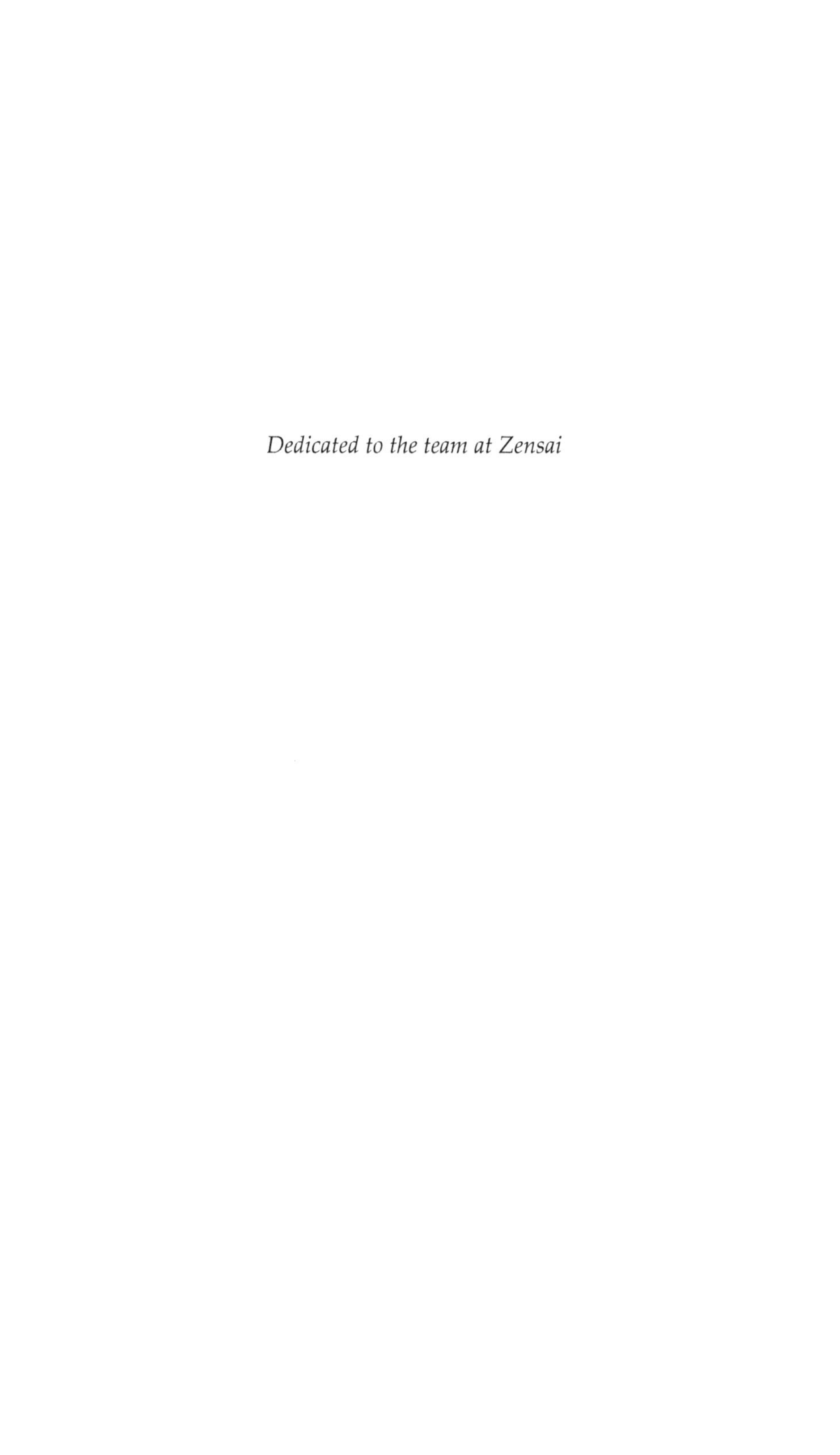

Dedicated to the team at Zensai

PART ONE
UNDER PRESSURE

CHAPTER 1
LOST

Evan wrinkled his brow. He cursed as he slammed his laptop shut.

On the other side of the marble coffee table, Rasmus Holst wondered what his friend had just seen in his email. He took a sip from his cappuccino and waited.

Holst, a tall man with angular features and a regal shock of dark hair that looked like a crown, was an unusual CEO. He was someone who could listen to anyone, anywhere, on any topic - and no matter what they said, Rasmus did not flinch. He listened. Perhaps it was his Danish upbringing, having been raised in Copenhagen, where he learned that decisive action can be a burden, not a privilege. He was not afraid to move quickly, but he knew that - sometimes - fools rush in. From catastrophe to conquest, observation was the first step.

Often, he would say he wasn't the smartest person in the room. But in any situation, he could identify who was. And then, if it made business sense, he would figure out how to hire that person. Even when being decisive, he always acknowledged his team's expertise at the outset. This approach preserved *tryghed* (Danish for a sense of security) - and allowed him to gather multiple viewpoints before offering his own.

Rasmus's instincts were impeccable. His Scandinavian resolve and business intuition made him lean in on trust, with a strong desire for a minimalistic org chart. His global perspective made him ambitious, but in a way that was collaborative - not aggressive or self-serving. That shared desire for success, perhaps the hallmark of the best CEOs, connected the two friends. So Evan, the CEO of a California-based company with strong ties to Asia, trusted him with what he was about to say.

Amidst the clinking of cups and clicks on keyboards within the crisp and modern café, the two men looked into the future. And told secrets that no one else was ever going to hear. Just before the pandemic, the west side of Manhattan underwent a bit of a transformation. A new hotel opened next to the Shed, a multi-purpose performance space, art house, and cultural center, unlike any other building on the planet. The Highline, a nearly two-mile long public park built on a historic elevated freight rail line, bordered Hudson Yards, the revitalized neighborhood that contained the Shed, the Equinox Hotel, and other landmarks (like the Vessel) near a high-end shopping complex called the Westfield World Trade Center. Near this impressive combination of buildings and people, in the shadow of Hudson Yards, Rasmus sat.

He sat with his cappuccino and his friend. And he listened. Just like those days in college together, when the two men unpacked the problems of the world. He listened.

"What am I not seeing?" Evan blurted out. He shoved his laptop into a backpack near his left foot.

These two CEOs were headed into battle together. They just didn't know it yet.

Rasmus wore a black t-shirt, dark tan tech pants, a digital watch, and a concerned look on his face. Evan Sato, a thin Asian man of average height but above-average ideas, wore jeans and a cotton sport coat. He hadn't shaved in three days.

Evan looked around the cafe. A media favorite with camera-ready looks and a propensity for cutting through the bullshit, Evan often appeared on the BBC, Bloomberg Europe, and multiple media outlets in the USA to talk about doing business in Asia. Rasmus noticed that his determined confidence in front of the cameras - and on LinkedIn - was flickering, like the light through the large picture window behind him.

Evan kicked his backpack under the table. He decided to say the quiet part out loud.

"How do I build a sustainable business in the age of AI?"

The AI problem was flashing on everyone's screen, everywhere, but nobody had solved it yet. What was the "right" governance model within this brave new world? Every CEO was supposed to have an answer. No one really did. Looking at the black box of artificial intelligence, Evan needed more than theories or trends for his business.

"I need an operating model that makes sense," Evan said.

Rasmus was puzzled. He knew that Evan had an operating model. A sustainable business. A track record of success. What had happened to throw his friend off balance?

Despite his public advocacy for AI, Evan's revenues had stagnated over the last twelve months. His sales were in decline. His profit margins were shrinking (due to a blowup in engineering and a missed product launch last spring). As CEO, Evan was doing something he had never done before: he was failing to meet expectations.

Transitioning to AI was more than just a technological initiative. It was a challenge to Evan's identity. Who was he, this CEO with multiple exits, if not a successful, determined and resolute leader? How could he advocate for AI when the tech didn't love him back?

For Evan, AI was an offscreen enemy. Beyond the increased efficiency, LLMs threatened his company, his career, his track record. Just like artificial intelligence was doing to everyone else on the planet, tech was forcing an uncomfortable change. And his Board of Directors was not happy with the plan.

A board member had confronted Evan - an ambush, really - in their recent Zoom call. What they had asked was impossible. Unthinkable. Evan rejected the Board's proposed direction - calling it an "overreach". He explained the situation to the tall Danish gent on the other side of the table.

The Board wanted to know: if the company was utilizing and advocating for AI, why didn't Evan's numbers match the competition - and the market? And the implied but unstated inquiry on everyone's mind but no one's lips: was Evan making the right decisions?

In the fateful call, Evan explained, he announced his plans for the company. The story fell on deaf ears. The Board quickly and unanimously rejected his direction and inserted their own. The positive sentiment he had experienced in private conversations was gone in the public forum of the conference call. That reversal meant that there were conversations Evan was not hearing. Sentiments he did not sense.

Secrets in the Slack channel.

Evan feared the unknown.

Evan's vision was not endorsed by his top investors. That lack of endorsement was a problem. Not every idea was a winner, he knew, but a vote of no-confidence was a challenge. Instead of heeding their direction, and folding up like a lawn chair, Evan dug in. He pushed back.

He doubled down on his own initiatives.

He was gambling with the people inside his company. And his Board didn't think it was a sure bet.

Rasmus listened with deep interest. "Any executive in tech, or marketing, or pharmaceuticals - even somebody that owns a manufacturing business in Missouri - you've got to be asking the same question: 'what's the right way to engineer this company around artificial intelligence?'" Rasmus stated.

"How do we create a competitive model against AI-native competitors?" Evan continued. That was exactly what his newest investor had said on the call just before he rejected all the CEO's ideas.

"There's no rulebook, no precedent, no map," Evan said.

"Of course," Rasmus reasoned, leaning forward, "you don't need a map if you have a GPS. If you have a way to navigate that meets you where you are, in real time, you're always up to date. Get lost with a map and you're in a world of hurt, 'cause you don't know where you are. Get lost with a GPS and it reroutes. Automatically."

Evan nodded.

Rasmus went on, "When it comes to governance and AI, people want a case study. Good luck. It doesn't exist. We are in the middle of the experiment right now. We are all creating the AI map from inside the journey."

The Dane picked up his cappuccino and looked at Evan. "People always want rules instead of tools."

Maybe that was part of Evan's frustration. Could Rasmus offer a solution?

Looking across the table at his friend, Rasmus wondered if Evan's Board was searching for a new CEO. His instincts were spot on - the search was already underway. Evan was in trouble, and both men knew it.

Luckily, Evan was not alone.

The Vessel, in Hudson Yards.

———

The Personalized Playbook

If AI can manage tasks, what is the indispensable role of the human leader?

After each chapter, a summary of key themes, with questions to consider, can help to align and internalize what's unfolding. More importantly, the Personalized Playbook summary is designed to activate your own expertise and insight - so that this book becomes your personal journey, as well as a tool to share with your team(s). Because stories without application are just entertainment. What is the story that you are creating, for yourself and your organization, in the age of AI? Evan is up against a lot. Rasmus is there to help.

Who's in your corner, supporting your leadership with a fresh perspective? The Personalized Playbook is a way to access your own insights and instincts, as you reflect on your own story.

Consider the following:

1. **Examine Assumptions:** What fixed assumptions does your organization hold onto, regarding AI? Are you working with a map, or a GPS, when it comes to decisions around human / AI interaction?
2. **Identifying with Identity:** As a leader, can you relate to the way Evan sees a slip in performance as a challenge to his identity? How is AI impacting your identity, as a leader and contributor to the

organization: is it helping, or hindering, your progress?

3. **Changing the Conversation:** What are the unheard conversations within your organization - maybe even within your Board of Directors - around AI adoption, and the role of team members in that equation?

4. **Facilitate Communication:** What are three things you could do to encourage a safe space for secrets to be shared before they become a crisis? What would happen if you had a more candid and open dialogue with the people that matter most, around AI and its effects on the team?

CHAPTER 2

THE SILVER BULLET

"When it comes to the game of AI, the rules have changed. The tools are all we have." Rasmus leaned into the small marble table as he spoke. "Ultimately, we are all finding our own answers in real-time. Like using a GPS."

Evan raised his eyebrows. The idea of navigation was intriguing.

Navigation was what he needed.

Encouraged by his reaction, Rasmus continued, "That GPS is what we are working on at my company," he said, tasting his cappuccino one last time before it got cold. "We are rerouting, right now - and helping our clients to do the same."

"Rasmus," Evan said, putting his elbows on the table, "I'm sitting at my desk in LA, breathing hard after another board meeting, wondering what the hell just happened."

Rasmus nodded. He knew the feeling exactly.

Evan continued describing the memory: "My CTO just pitched another AI investment. The CFO told everyone our margins were flat. The CSO is doing everything except a handstand to drive numbers. My CHRO just looks overwhelmed because no one cares about attrition. All I know is this uncomfortable truth: *we are not ready.*"

Rasmus offered a detailed scientific rebuttal which he engineered with great care:

"No shit," he replied.

Evan laughed in spite of himself.

Evan lifted his hands as he spoke. "The headlines today talked of another AI-native company that reached $100 million in revenue in less than a year. Less than a year!"

Rasmus knew of the company. The vibe coding legend was yet another AI success story: a unicorn with a billion-dollar valuation. And in an adjacent business sector, no less. So now Evan was facing a competitor with more traction, less people, better margins, and greater investor enthusiasm.

Evan's board wanted to know what he was going to do about it. So did he. Make no mistake, Evan had a strategic plan and a vision for the company. But no one - including Evan - believed it would work.

Evan said, "In the past, you could run a company where your average revenues per employee were $200-300,000. And that was good, right? If your company was making $100 million, you had a payroll of 300 or 400 people. Not anymore! Now the expectation is to create an AI-native company, where you have one tenth of that headcount. You've gotta have revenues of $2-3 *million* per employee now—because others are doing that model with AI! OK, great, but what if you built your business in a more traditional way? You're running a company that is 10 or 20 or 50 years old. You are not AI native. You're trying to engage and utilize AI as a tool. Everybody is grabbing apps or building apps or using AI wherever they can, that's what we are trying to do. It's a totally different org structure and governance and…" his voice trailed off. "What about the team? What about the people?" He stopped.

His words cut so deeply that they finally hit the bone. People, Evan thought to himself, that's where AI stops. And where it all starts. People.

Evan looked at his old friend. He wanted to tell Rasmus of his plan.

But those details would have to wait. Because these next words spilled out of him like blood from a deep open wound.

"I can't just cut people! I've worked hard to build this team. We aren't AI-native, but we use AI, of course, and our products are great and AI-centric but that's not enough for the board and for our investors so I have hundreds of people on the payroll - just like you, Rasmus - and I can't wave a magic wand and just 'poof', transform my EBITDA!"

He was talking about earnings before interest, taxes, depreciation and amortization - EBITDA. A crucial number for a CEO. For many investors, the only number that really mattered.

Evan's number was not good. It was getting worse. Something had to be done.

So he did what he always did: he took bold and decisive action. His counter-intuitive plan had shocked his investors and board members.

He didn't care.

Or did he?

"EBITDA is the holy grail, you know this, and all these investor terms like Rule of 40 exist within a context and that context is the structure of the company - and the people inside it," he said, finding his thoughts by saying what the two men already knew. "I'm wondering about any kinds of layoffs and what that is going to do to morale because the minute my best people see a large-scale reduction, they are going to go else-where. But they are already seeing the shifts. We are in the middle of the change, not in front of it! People want case studies and reference points-ha! That is hilarious - because there's never been anything exactly like this. There is no clear perspective, no frame of reference, but people try to find it. 'AI is sort of like a calculator' is like saying 'COVID is like a cold'. Similar, maybe - but nowhere near the same. With wildly different repercussions!" Evan realized he was raising his voice and caught himself. He took a breath to gather his thoughts.

The next words were conspiratorial - almost a whisper: "People are already talking about how companies are chang-ing. The writing is on the wall," Evan said.

"What kind of writing? What do you mean?" Rasmus asked.

Evan listed several famous-name companies that had laid off tens of thousands of workers despite record profits.

"You do your corporate hygiene," he said, referencing how small reductions in the workforce can make the balance sheet look more attractive and tidier. "You cut your bottom five percent; it gives you a better annual report - everyone knows this. People are seeing that layers - *layers* - are being cut off. That's not the bottom 5% at the end of the year. The Fortune 500 is rethinking the workforce. Tech companies are leading the way, but others are following. They have to! Manufacturing, healthcare… My board is looking at EBITDA and telling me to 'fix it'… fix the number…" He shook his head as he repeated himself. "The number…"

Evan sighed.

He said, "I can't tell if AI is the greatest tool in the history of business or if there's a werewolf loose in our headquarters."

That was a weird reference, Rasmus thought to himself.

He was trying to remember how to kill a werewolf. Was it a wooden stake?

Evan sat back in his chair and kicked his backpack under the table once again for good measure. "Where is the model that tells me how to survive this, without killing what's still working? I'm not afraid to make cuts but I'd like to do it without destroying trust, or culture, or momentum."

"You want a silver bullet for this werewolf?" Rasmus said with a smile. Evidently, he did remember.

"Unfortunately, I don't have a silver bullet," he continued. "But AI is not your enemy - it's not a monster that's on the loose.

"I do have a team that is helping me. And helping our customers and our investors with these exact same questions. And a few others that you haven't mentioned yet. But I know you're thinking about them - because we all are," Rasmus said.

He explained that his team was creating a kind of new positioning system - a sustainable governance mechanism for the age of AI. He said it wasn't just some newfangled strategy or retrofitted set of soft skills for HR. It was a way to change the conversation in the boardroom. And the breakroom. And everywhere in between.

Then he said what Evan really needed to hear: "It helps you to fix the number."

Evan wondered to himself, which one?

Reading his friend's mind, Rasmus said, "All of them."

A bold claim, Evan thought. Would it work for him? At this point he needed to try something. Anything. He needed a fresh perspective. Could Rasmus offer some kind of guidance that would turn his plan around?

Someone close to Evan, on his Board, wanted him replaced. Evan's bold moves had triggered a troubling clause inside his employment contract. The clock was ticking on Evan's career. "Let's walk back to the hotel and meet my team," Rasmus said. "We'll tell you everything we have discovered so far."

The Personalized Playbook

1. **The New Success Metric:** Evan highlights the new expectation of $2–3 million in revenue per employee driven by AI-native competitors. What is your organization's current Revenue Per Employee (RPE)? How does this number compare to AI-native leaders in your industry? What specific gap must you close to compete in your industry, and to align with this new standard?

2. **A Werewolf, or an Asset?** Evan struggles to distinguish AI as a "greatest tool" from a destructive "werewolf". Where in your organization are you applying AI strategically as a tool? Where is the fear of AI (the "werewolf") driving reactive, fear-based decisions like management by headline or destructive layoffs?

3. **The Cost of Corporate Hygiene:** Evan worries that large-scale reductions will cause his best people to leave, sacrificing momentum and trust for a short-term boost in EBITDA. Beyond the immediate cost savings, what is the calculated cost to your culture and talent pipeline if you were to mandate a significant workforce reduction purely to "fix the number"? What is the counter-intuitive bold action (like Evan's first move) that prioritizes long-term trust over short-term financial optics?

4. **Beyond the Silver Bullet:** Rasmus claims his approach can help "fix the number," meaning "all of them". If you stopped searching for a single "silver bullet" solution, what three organizational numbers (e.g., EBITDA, RPE, Employee Engagement Score,

Product Innovation Speed) would you prioritize fixing simultaneously to signal a balanced, human-centered approach to AI integration?

CHAPTER 3

WIZARDS AND WARRIORS

Nina and Robin entered the chat.

On the short walk from the coffee shop to the hotel at the center of Hudson Yards, Rasmus made a quick phone call to his two chief lieutenants. He explained the situation and told them how to prepare for the upcoming conversation. Rejoining Evan after the call, they walked quickly to the hotel and boarded the elevator in the lobby.

The elevator doors opened, and the two men stepped into the restaurant on the 24th floor.

Evan was preoccupied. He made lists in his head. He always had, and today was no different. His attention to detail made him replay last week's Board meeting. What am I missing, he wondered to himself - as he watched mental game films, trying to find his blind spot. What is it that I'm not seeing? The investor triad - there were three seats taken by his lead investment firm - was making a ballsy move. Or creating a clear and present danger to the company. Was this a test?

The men exited the dining area to discover a large outside

patio - an enormous space where 200 people could enjoy an uncrowded cocktail party, surrounded by skyscrapers and views of the Hudson River.

The 24th Floor Patio

A large mid-afternoon shadow sliced a diagonal through the space, blocking the sun for those who stayed close to the building while helping those near the bar to work on a tan.

Tucked into the shadow and seated in a low-slung wooden chair, a tall woman with sandy blonde hair was listening to an energetic guy wearing bright orange running shoes. The woman started to laugh as the man stood up to make a point. They were both smiling when they turned to see Evan and Rasmus.

"Evan! Wow!" Robin Daniels said, extending his right hand.

Nina Carøe stood up to exchange greetings as well. She said, "Why am I not surprised that you two know each other? I guess it's true. Everybody knows Robin.

"Hi Evan, I'm Nina Carøe." She pronounced the o-slash like a standard "o", "care-oh", to keep it simple. Like Rasmus, Nina worked at the company's headquarters in Copenhagen.

Meanwhile, Evan and Robin played a quick game of connect the dots. After 25+ years in Silicon Valley, working at companies like Salesforce, LinkedIn, WeWork and Box, Robin Daniels was a Silicon Valley warrior, and he had met Evan on more than one occasion. Evan remembered that Robin had been a part of 2 ½ IPOs. Evan was an early investor in Matterport - one of Robin's successful exits - and they chuckled over the "½" part. WeWork was a bit of a mixed bag for owners and investors alike, they agreed. Silicon Valley was like a small town - everybody seemed to know somebody.

Like Nina and Rasmus, Robin was a native of Copenhagen - but he had moved to the US in his early twenties. Once upon a time, he bought a one-way ticket to San Francisco, launching a career that would take him all over the world.

Although Evan knew of Robin's corporate history and reputation, he didn't know that Robin and Rasmus worked together. Rasmus explained that, in a few hours, Robin would deliver the kickoff keynote for the conference. Evan congratulated Robin, saying, "I guess that explains why your smiling face is on all the TV screens in the hotel." Evan was planning on being at the presentation, and reconnecting with his own team at the kickoff. Then he confessed that he didn't know much about the software company that his friend Rasmus was running.

Rasmus, Robin and Nina took turns filling in the gaps to get Evan up to speed.

The company's software was the only cloud-based AI-powered learning, performance and engagement platform built into Microsoft® 365. Their customers covered multiple sectors - manufacturing, healthcare, financial services, hospitality, and more. What started as a learning management system had become something much more than just a repository for training programs. Rasmus explained that the heart of the expansion was an acquisition of a Danish company called ValueBeat. The co-founder of that company was Nina Carøe.

"I told Nina that I loved what she was doing at ValueBeat," Rasmus said, describing the purchase. "Because she was data-driven and her solution was data-driven around the people stuff. So, we made plans to buy the company."

"And then in that meeting," Nina said, with emphasis, "I was wondering, 'why is he asking all of these questions about me'?" That remark made Robin laugh.

"He wanted to hire you, of course," Evan said, smiling.

"He knew what I cared about: people," Nina said. Rasmus chimed in: "That's right!"

"Nina is our Chief Human Success Officer," Robin added.

"Excuse me? She's what?" Evan said.

Robin pointed at the chairs to Evan's left. "Let's sit down and we can explain."

Nina picked up on Evan's question. "Human Success is a framework, a platform and a philosophy. It's something that we've designed - the three of us - to help leaders to lead from a more intelligent place. Human success informs every team, every collaboration, every decision. And it allows for a numbers-based insight into performance."

"At the company level, or an individual level?" Evan asked.

"Both," Nina said.

"I guess I'm the first Chief Human Success Officer," she said with a smile. "Our company is helping leaders and employees to track progress in a new way, with AI at the center of the conversation. But we never forget that the outcome is always Human Success," she explained.

"You know how inputs plus outputs equals outcomes, right? Pretty straightforward, I guess. I share this in all my keynotes," Robin stated. "Lots of folks treat AI like it's the outcome, but it's not. It's just another input. Which means that, if we structure things correctly, it's something that is absolutely within our control."

Evan took this in. He nodded.

"So, for us, we are a Human Success company," Rasmus said. "Saying 'human resources' or 'HR' treats employees like numbers. Like factors of production, or PPE," he said, abbreviating for property, plant and equipment. "The best part is, when you have the elements of Human Success in place, you can speak about your people and your performance with a level of analysis that rivals the financial details of your CFO," he said.

Evan thought that sounded like a bold claim. But he stayed silent - and curious.

"Every Chief Human Resources Officer wants a seat at the table, and wants to be taken seriously," Nina shared. "But without hard numbers around performance, career paths, collaboration and more, the conversation feels…soft. Squishy. That's why I left HR and started ValueBeat. HR was broken. HR was tracking absence. Human Success tracks impact. Ultimately, HR wasn't serving employees. It wasn't serving the leadership. And it wasn't possible to have a concrete and meaningful conversation with investors," she said.

"Until now," Robin added.

He continued, "A team is really just a collection of people trying to achieve great things together. And now, AI is part of that team. Together, every business wants the same thing. The outcome we are all working towards is Human Success."

"Sounds like it's a new method of governance or structure," Evan offered.

"In simple terms," Rasmus said with a nod.

"Like a GPS?" Evan queried.

Rasmus nodded, but a little more slowly this time – smiling as if to say, "Now you get it."

"Human Success takes in the entire organization and provides a numbers-based analysis for performance. It's both a dashboard view, and an individual roadmap, that rivals the level of detail that your CFO sees on the balance sheet."

Evan was intrigued, and he wondered if this Human Success idea would fit for his company - and what that might look like. As he considered the possibilities, Nina shifted the conversation. "Evan, can we talk about what's going on with you?"

Robin added, "Rasmus said you are getting some pushback from your board. And there are some questions about AI?" he said. "What's up?"

As Robin and Nina finished each other's sentences, Rasmus leaned back and signaled a server.

At the coffee shop, Rasmus's spider-sense was tingling as he spoke with his friend. Something unknown - and potentially dangerous - was happening around Evan. Rasmus wondered how the investment profile for Evan's company had changed. What exactly was going on?

His instincts told him that Evan was facing an activist investor - somebody who had heard that AI plus cuts equals growth.

Business, Rasmus knew, was never that simple. Even with AI in the equation.

An activist investor can seem bombastic and challenging. But when you hear what they're about, if you really listen to what's going on, even the most incompatible person somehow has to admit something very difficult: "Well, they aren't entirely wrong." The key, with an activist investor, is to defend your position without getting defensive. Stick to your guns, without dropping your guard. Oh, and stay open to new ideas.

It wasn't easy, Rasmus realized.

Outright rejection of an Executive's plan is a blunt instrument and rarely used. The whole plan was bad, they had said.

Really? Rasmus wondered to himself: no part was worth keeping? Hard to believe. But that was what the Board had wanted. Suggested. And then, demanded.

Rasmus was a big fan of J.R.R. Tolkien, the author of the *Lord of the Rings* trilogy. He wondered if artificial intelligence was a version of Gollum's ring, in both its allure and its mysterious power. Tech had found its "precious" - a magical and misunderstood charm that could influence even the most intelligent investors to over-pay for assets and over-index on cutting headcount. Or perhaps enable these same folks to accurately predict the future and make billions in the process?

Had AI exercised some kind of *Lord of the Rings* dominion over Evan's Board? And was someone using the ring for invisibility right now - hiding his or her true intentions from the CEO?

Somebody with a blunt instrument wanted to cut into Evan. On the surface, it seemed the board didn't understand AI, or how to manage its impact. But that lack of understanding didn't inhibit their confidence - or their demands. Rasmus sensed it. So did his friend. And soon, so would Robin and Nina.

The real game was about to begin.

———

The Personalized Playbook

Why a Fresh Perspective Matters. Here, human success reframes the conversation around people from "factors of production" (HR) to a measurable, numbers-based performance metric that leaders can confidently present to the Board.

1. **AI: Input or Outcome?** Robin states that AI is just another input within the organization, and Human Success is the desired outcome. How is your leadership team currently talking about AI? Are you treating AI as an input (a tool you control) or are you already treating AI as the *desired outcome* (a force that must be accommodated)? How does this distinction change your strategic discussions?

2. **From HR to Human Success:** Nina's experience is that HR was broken because it lacked the hard numbers necessary to engage leadership and investors. Do you agree with that assessment? What three performance metrics related to your employees (e.g., collaboration, career path progression, retention of high-potential talent, etc.) would you need to measure and quantify to make your "people conversations" as concrete and influential as your CFO's balance sheet report?

3. **Rasmus wonders** if the mystique of AI is creating a "Gollum's ring" effect on leadership—causing smart investors to over-pay for assets or over-index on cutting headcount. Where might your organization be over-indexing on the *magic* of AI (e.g., aggressive staffing cuts, chasing flashy tech) without fully understanding its impact on core human values like trust and culture?

THE NUMBERS GAME

"40 percent," Evan said. He shook his head. "They want a 40 percent cut."

Evan was explaining what happened on the call with the Board. Evan had offered his strategic plan, via screen share. It was the same one he had discussed in private with pretty much everybody (he thought). The plan was aggressive, innovative, and comprehensive. Until it wasn't.

That was when a new voice from Arbor Partners joined the Teams call. Arbor Partners was Evan's lead investor, holding three of the seven seats on the board.

A man with a brogue joined the conversation - a new and strange voice that captivated attention. And curiosity. "Evan, there's a deep need for more cuts," the new voice said.

On the screen, the conference room at Arbor Partners showed two of Evan's three investor board members sitting in silence. Who was speaking?

Eileen Wilcox, a fixture and fan of Evan's from the investment firm, was the third voting member from Arbor.

Evan thought Eileen was there, via a black box on the screen that was simply labeled "Arbor 3".

He was mistaken. He did not know who was behind door number three. Because a man's voice was speaking.

One outspoken board member was swinging his… well, his weight… around.

"We've been watching and waiting for top line growth but now it's slowed. I dinna ken what's going on here, but things have got to change. We need you to look at EBITDA," the mystery man continued, revealing a bit of a brogue which took a moment for Evan to wrap his head around. "Tha' plan is shoogly. You gotta cut the headcount. We're all agreed here at Arbor. Put it to a vote. But here's tha' deal: you've got to cut 40% of your workforce. 40 percent." The voice was firm and resolute. "That's the number. Other companies are ahead of you, with AI. Less people, Evan. You've had two and a half years here. It's time to get to work."

Evan saw the guys at Arbor Capital nodding silently in the conference room. He glanced at the square at the bottom of the screen. Board member Zahra Al-Mansour leaned back in her Miller Knoll chair. Her face showed her agreement as well.

Arbor's three votes, plus Zahra. Four out of seven was a majority. Someone at Arbor had staged a coup.

"We're gonna need a new plan," the voice said. "This one won't work. There's no need to debate. Just come back with a new one, where you show tha' reduction in headcount."

Robin was curious. "So, what did you do?"

Evan had been ambushed before. He knew how to pivot - and how to isolate risk.

Evan spoke of his next move on the Zoom call. "'Tell me what your outcome is. And what is the timing of your outcome? I will come back with a plan that meets your needs.' That's what I said."

Rasmus nodded. He would have played it the exact same way. Where was this 40% number coming from?

That was when the Arbor 3 screen changed. Someone turned on their camera.

"The mystery investor," Evan explained, "was Lachlan Moore."

"Who?" Nina asked.

Lachlan, Evan explained, was the powerhouse billionaire investor that had been hiding in the tall grass, like a snake about to strike. He was the Scottish hammer that had smashed more than one underperforming investment in his portfolio.

On the screen, Evan could see that Lachlan sat in an office that would have made sense in the nineties. Dark cherry wood. No screens on his desk. Industry trophies on the shelves. No pictures of family. No poses with his teams or other corporate leaders.

Pushing 70, his dark eyes were intense. Hungry. Always looking for opportunities. In an old-school burgundy executive chair, Lachlan was a grumpy grandpa with a brogue. A lone wolf. And an unknown quantity for Evan.

An unexpected visitor.

That was because Lachlan had made his money in an unfamiliar sector: oil and gas. Lachlan Moore, Evan explained, was the most powerful energy investor ever to come from Aberdeen. At one point, Lachlan was one of the wealthiest men in the UK. Nearly a decade ago, he left Scotland and shifted his attention to the United States. He moved to Houston, Texas (the energy capital) to be closer to his investments.

Last year, he leveraged his fortune and connections to secure his Senior Partner spot within Arbor Capital. The Silicon Valley VC firm seemed a strange spot for a Scottish oilman. But their portfolio was a great place to swing a hammer and break things. His California counterparts considered him a bit of a cowboy, Evan explained. The conservative vibes of Texas, combined with the State's tax structure, appealed to him - and confounded his other partners.

Lachlan had convinced Arbor Partners that he would help

expand their portfolio in new markets. True to form, he quickly stepped out of his lane once he was on the inside at the firm. He began making his own rules. His partners from Aberdeen used one word to describe him: *gallus*. Indeed, his moves were filled with swagger - and he never asked for permission when he wanted something done.

Lachlan thought he understood tech and AI. That's what his ego told him, and he believed it. His track record and bank account left him with a flawed logic that he always knew best. His ex-wives disagreed. But if he had listened to either of them, he would still be married. And on speaking terms with two of his three kids.

Rasmus, Robin, Nina and Evan had seen it time and time again: wealth can be blinding.

But try telling that to a billionaire.

Lachlan's financial position within the firm enhanced his swagger. Step one was getting Eileen Wilcox to step back and work on a challenging new energy project. She agreed. With Eileen off the chessboard, the junior board members from Arbor were pawns to persuade.

AI, for Lachlan, was simply the next logical step in automation.

In the same way his companies provided innovative depth analysis to oil drilling - doing what humans never could or would - AI was going to do the same. Only this time it would be Texas-style, which is to say bigger. But not always brighter. Or better.

Lachlan's approach was leadership via headline. He

bought into the AI publicity that high-profile CEOs like Marc Benioff had popularized. Robin's old boss at Salesforce had made news every time he announced a layoff, slicing headcount and either embracing or rejecting agentic AI - it was up to his stockholders to decide which way was best. Nobody, not even Benioff, could predict the future. But that didn't stop Lachlan from trying to shape it. With a hammer.

Raised with an old-school playbook of slash and burn, combined with a cursory knowledge of AI, Lachlan Moore offered a binary choice for Evan. Cut or be cut. Perform, or move out.

Lachlan reasoned that, if an algorithm can write a report, the person who used to write that report is now redundant. "A task is either done by a human or by a machine," he told anyone at Arbor Partners who would listen.

Lachlan liked to hit first, and he came out swinging on the board call. But he was not nuanced. Leadership skills that required elegance and dexterity were beyond his reach. In contrast, those precise maneuvers were Evan Sato's superpowers.

The Scotsman couldn't see that his conclusions about the new AI paradigm were superficial and uninformed. "Because inside investment circles," Evan explained, "you live by the golden rule."

"What do you mean by that?" Nina asked.

"Whoever has the gold makes the rules," Evan stated.

As Evan spoke, Rasmus began running scenarios in his head. He knew that headcount is only one way (and rarely the best or first way) to find savings or efficiency. Rasmus was curious. "So did you offer them a sacrificial lamb - some other way to cut costs - on the call?"

Evan did. He would change some distributor commissions. End a consulting contract. Cut a small division entirely. The board was satisfied. For now.

Evan pushed back on the headcount thing, the 40 percent, trying to discover where it was coming from. "That number felt way too high to me," he shared with his friends on the hotel patio.

"I hear ya," Robin said. "I'm wondering: does your board want to gut the company and sell it?"

Nina said, "Are there some financial problems you aren't telling us? Are you guys an acquisition target?"

Evan shook his head. "No," he said. "Our growth has slowed. But we have good prospects, good internals, good potential."

"So," Rasmus interjected, "when you told them you weren't comfortable with the number, what did Lachlan say?"

"He has profit objectives around EBITDA that they need me to hit. They want us to use AI even more than we are now. He explained their expectations, which was a good thing. I see what they want. It's a stretch. But it's not insane.

"And then he said that if I didn't like his number that I should come up with my own," Evan relayed.

"Your own reduction in headcount?" Rasmus asked.

"Yes," Evan said.

"That was the guy's final question to me. 'What's the number, Evan? You tell us: What's the number?'"

CHAPTER 5
EVAN

Evan Sato feared the unknown.

It was a fear he knew well.

When the cancer came for his mother, he was not ready. No one was. The attack was swift. His family fought back. But there are some battles that no one can win.

Emiko Sato, Evan's mother, was an artist. She was a practitioner of *ikebana*, the Japanese art of flower arranging. Her oldest son, Yee-Un (that was Evan's given name in Japanese), was her treasure. "*Musuko yo, kokoro ni kimeta kotonara nandemodekiru*," she would say. "You can do anything you set your mind to, my son."

She celebrated his schoolwork. Encouraged him to help around the house. Taught him to be kind to his younger sisters.

The diagnosis came when Evan was a teenager, living in Honolulu. Mom was stumbling. Weak. It was a rare form of cancer. She was 41.

A new experimental treatment, called "proton therapy", was the best possible solution. Proton therapy, at this time, was in clinical trials. The remedy was available in only one place.

Evan's family moved from Honolulu to San Bernardino, California.

Evan's father, Hiroshi, was a master carpenter, a *daiku* from Japan, who emigrated to Hawaii and became a US citizen. He was a craftsman with meticulous attention to detail and efficiency. He ran a factory in Oahu that produced curios and souvenir items for the Hawaiian tourist trade—an extension of his woodworking skills and his American entrepreneurial spirit.

"Which of these tools," Hiroshi had asked a ten-year-old Evan, "is the most important in this wood shop?" Evan looked at the chisels – a set of 10, each one slightly smaller than the one next to it – hanging uniformly on the wall. The shelves were filled with cardboard boxes, some containing drills, sanders and other handheld tools. Glass jars filled with screws and nails sat next to the boxes. Evan thought about how nails are the connectors between one piece of wood and the next. Could that be the most important tool?

Evan looked at the large equipment in the woodshed. There was an 8-inch long bed jointer: a long, low heavy table with a spinning cutter head in the middle. It sat in the middle of the room as a sign of its importance. In the corner, a Delta Rockwell band saw: a five-foot tall column of grey painted steel that his father called the "curve-maker". Without it, forming elegant architectural details would be nearly impossible.

Evan's father took his son's hands in his. *"Anata no te,"* he said. Evan knew it instantly: *your hands.*

"These are the most important tools in this shed." Evan looked down at his thin fingers. Beyond his upturned palms, the hands of a 10-year-old boy, he felt his father's fingers. His father's hands were supporting his own.

In his home woodworking shop, Hiroshi taught Evan to measure twice and cut once.

"The way you set the saw," Hiroshi would say, in perfect Japanese, *"determines the cut."* He reinforced that the hands of the carpenter were the key – because any tool is only as useful as the skills of the one who wields it.

Evan lifted his left hand and rubbed the scar on his left cheek. The scar, from a stray shard that rocketed out of an old piece of wood, nearly cost him an eye. It was a reminder to pay attention. Watch out for hidden dangers.

Carelessness had consequences, Evan's father said.

Evan's jaw clenched at the thought. The scar was the result of a freak woodworking accident. Even when you are paying attention, accidents can happen. The unknown always awaits.

Evan's mother had not been careless. And neither had he.

Especially on that day.

Evan held his mother's hand. He supported her thin fingers as his father had once supported his, when he was just a child.

Her eyes blinked at her oldest son, who was now learning to drive – a young man on the edge of 16. As she lay in the bed connected to the machines that were helping her to stay alive, she smiled as she spoke.

"Go take care of your sisters," she said to him. "They need you." Evan kissed her forehead.

"You can count on me, *okaasan*." He looked at her eyes. "I love you, mom." Evan grabbed his backpack off of the floor, near the bedside, and headed for doorway. "Time for school! Let's go!" he shouted into the hall.

When he came home that afternoon, his father was waiting. He met him at the front door with a warning. "Yee-Un, stop."

Hiroshi's hands were on Evan's shoulders. That was so he

could hold his son. Because in three seconds Evan would crumple to the floor.

"Yee-un," Hiroshi said, trying to find the words. "She's gone."

They fell to the floor together. They cried. The two Sato men were broken, like old wooden chairs that couldn't take the weight placed upon them.

The girls - Evan's sisters - would be coming home from the bus stop soon. Practicality took over for the tears.

Evan dried his eyes. He honored his mother's wishes. He helped his sisters. He helped his father. Evan dealt with medical bills - because, even though his mother was gone, the bills stayed alive.

Evan had to assist his father with the finances, by making sure the bills were paid. Everywhere, Evan helped.

In the days before zoom and widespread internet, Hiroshi traveled to Honolulu. A lot. He didn't want to, but medical bills for an experimental cancer treatment meant he needed money. He needed to manage his business. And that business was 2,578 miles away.

Hiroshi felt he was never enough.

As a result, Evan' high school experience was anything but typical. He learned to braid hair. He watched silly anime shows. He arranged playdates. Evan filled the gaps.

Evan always kept things moving when his father was in Hawaii. He did what needed to be done. As he worked with his father, Hiroshi would say in Japanese: *"Waste not a single moment. For a thousand small inefficiencies can become one large failure."*

Evan knew how precious those moments could be.

The Japanese concept of enduring the difficult or even unbearable burdens of life with patience, self-denial, and dignity is called *gaman*. The word came to define Evan's approach to life, after his mother's death. He was stoic. He was efficient. He watched out for hidden dangers. Evan knew better than most that you can't prepare for everything. The unknown was always there.

Evan was ambushed in the Board meeting. It was a surprise attack. Like his mother's cancer, the conference call was the threat he never saw coming.

Evan learned long ago that he could not afford to be paralyzed by fear. He had survived the loss of what he loved the most. An unseen enemy with uncertain power and influence had taken his mother. Now, he was determined that a surprise attack would not take his company. Or his career.

He rubbed his chin. "If Lachlan wants to cut 40% of the company, he can start with me," Evan decided. That was easy to say, but hard to do.

He thought about his three-year employment agreement.

Leaving now meant leaving a lot of money on the table. A. Lot. Of. Money.

Luckily, Evan had something that's worth more than money: a stellar reputation. He was an industry icon. His intellect, media visibility and a stellar track record were valuable assets. He could afford to walk away, even when every financial advisor would call it a mistake. If Evan Sato knew one thing about money, it was that he could always make more. Even after a loss.

44

The Board's ridiculous demands were too much. Cutting 40%? The treatment is worse than the disease, Evan reasoned. The cure would create complete system failure. He'd seen that before. Why wait for the inevitable? He started planning how to maximize his cash on exit. Evan was going to quit.

The Personalized Playbook

It's a common misconception that the past creates the future. The future always comes from one place and one place only: right now. The past reminds us, it doesn't define us. Otherwise, progress would never occur. We would be stuck in the past. Or imprisoned by it.

1. **Embracing the Past:** Considering Evan's history, look to your own. What is the personal, non-business experience (a trauma, a profound lesson, a failure) that is currently influencing how you perceive and react to the unknown challenges of AI and organizational transformation? Neuroscientist Carmen Simon, author of *Impossible to Ignore*, says that the past is only helpful in that it allows us to predict the future. How does the past allow you to do that?

2. **The Daiku Principle:** Evan's father, the master carpenter (*daiku*), taught him meticulous attention to detail. What is your fundamental, non-negotiable "Daiku Principle"—the core value or standard of craftsmanship—that guides how your business operates? Does your current organizational model support or contradict this principle?

3. **The Worse Cure:** Evan decided the demand to cut 40% was a "cure that would create complete system failure" - perhaps as he saw, firsthand, with the effects of the experimental cancer treatment. In your organization, what is a currently accepted "best practice" that, if pursued to its extreme, would prove to be worse than the problem it's trying to solve?

4. **The Value of Reputation:** Evan was willing to quit and leave a large financial settlement to protect his reputation and integrity. As a leader, what is the single non-financial asset (trust, reputation, culture, employee morale) that you would be willing to risk your financial security or position to defend? Have you clearly communicated the value of this asset to your Board and investors?

CHAPTER 6
A PEACE OFFERING

"So, this number is like net recurring revenue...but for your people?" Evan asked. He was listening to some encouraging news from Rasmus. Was there a possible solution here that didn't require him to look for a new company to run?

The conversation at the Equinox Hotel in New York City was getting interesting. The afternoon sun dipped towards the horizon and the shadows began to fill the entire patio. The kickoff presentation was drawing near; Robin excused himself to go do a sound check for his keynote.

Rasmus said, "Well, NRR - net recurring revenue - shows how customer value shrinks or grows over time, right? So, this score - this Human Success Score - shows how team capabilities and trust are either growing, or eroding."

Nina jumped in. "Old systems held people accountable. But now, we need to make manager impact measurable. We need a score that links people-sentiment and capability to financial outcomes. That way, we can track things like responsiveness, collaboration and burnout risk. Many organizations are cutting back on middle management," Evan nodded. That felt familiar to him. Nina continued, "We need a way to measure progress."

"But how do you ensure that 'collaboration' score isn't just measuring how many Slack messages they send? How do you measure true impact?" Evan wanted to know.

"Linking data to outcomes," Nina said, "is the key. The Human Success Score isn't an isolated number, it's a conglomeration - and measurements, like the number of Slack messages, aren't relevant to outcomes. If it takes 2 messages or 20, the question is: did we reach the goal? Did we hit the objective?"

Evan agreed: objectives were the key to turning data into insight. Measuring team outputs or even tracking minute inputs (like Slack channel messages) was an incomplete approach. Evan sat back in his chair.

Human Success would measure the quality of his management, not just the quantity of management maneuvers. Human Success would look at coaching, development and retention - not only for the organization, as a whole, but for each individual employee.

The sound of a helicopter at eye-level made Evan turn to his left. Patrolling the Hudson, on its way into New Jersey, the helicopter made a wide turn in front of Hudson Yards. He felt a little like the helicopter pilot – like he was starting to see things from a really interesting viewpoint.

Evan relied on AI, with apps on his phone and communication in his inbox. Things like creativity and decision-making were things he never wanted to outsource.

Was he being too precious about his own humanity, when it came to AI? Evan didn't think so.

Maybe he was over-estimating his own ability? After all, if AI could vibe code in seconds and access all of human history in whatever language you wanted in a moment, how could humans maintain a leadership role? Would AI, Evan wondered, come to replace him?

Maybe Lachlan had a point. Maybe Evan's hesitation and resistance to cutbacks was just delaying the inevitable. But his gut instinct told him that was not the case.

Evan recalled an AI-generated image where two people in the picture had six fingers on each hand.

While he was still CEO, he would do whatever it took to help the business succeed. Period. He would hire and fire as needed. But also, he was not going to perform surgery with a lawnmower.

He began to reflect on an interview he had seen with Louis Rosenberg, the co-author of *Our Next Reality*. The Stanford professor was considered by many to be the godfather of virtual reality, having designed the first VR system for the U.S. armed forces in the nineties. In recent years, Rosenberg shifted his gaze to artificial intelligence, founding a company called Unanimous.ai. Rosenberg turned his attention to something called Swarm technology as a model for AI implementations.

The power of the swarm, according to Rosenberg, can be observed in schools of fish, flocks of birds and (you guessed it) swarms of bees. The idea resonated with Evan. Basically, Rosenberg looked at nature to see how collaboration can make people more effective at decision-making. For example, an individual honeybee (with a brain that's the size of a grain of sand) can't really do much in the way of thinking. To make better decisions, the bee accesses the combined knowledge and wisdom of the hive via communication. Leveraging the experiences of others, bees become significantly "smarter": making better decisions, avoiding dangers, locating pollen, and in simplest terms: surviving. In the same way, human beings (with our advanced communication skills and networks) can leverage the natural swarm structure to advance the whole.

Evan instantly got the concept: if you collect data from a group, that approach will outperform the average person by a vast percentage. Adding AI into the mix amplifies the group result. But what's the right combination when AI is on the team? Is it collaboration... or capitulation (where everybody just assumes that AI can do it better, and lets the AI do all the work)?

Rosenberg turned to sports betting (a predictive model) to test the power of the swarm. He ran experiments on hockey scores, baseball scores, and the like. The group did well at predicting who was going to win, better than any individual could on their own. Then, he introduced AI.

Performance suffered. It did not improve. This outcome was unexpected for Rosenberg and his colleagues. What was going on?

The groups deferred to the chatbot. The humans outsourced their thinking to AI.

Assuming that artificial intelligence knows best, the study

participants let AI do the work. As a result, the predictions got worse. Not better.

However, when humans stopped deferring to AI - and started treating the chatbot as just another voice in the conversation - the numbers significantly improved.

It seems that the compute power of the large language model was accelerated by human intuition, human input, and real-world context. In other words, things that only a human could access were the real game changers. The key was working in tandem with AI.

Rosenberg explained it like this: If you're in a room with five other people, even if some of those people are really, really smart, you still would consider their opinions and viewpoints as one among many. After all, having a PhD in neuroscience doesn't mean you know who is going to win the game between the Bruins and the Maple Leafs.

When people overestimate AI, it makes the group less smart. Their predictions, using AI alone, were not as powerful as when everyone contributed - and didn't place unearned confidence on the AI.

When you are clear about what you know, and what you don't, you can calibrate your own level of confidence appropriately. You don't diminish or dismiss your own insights, or stop sharing your voice, just because you're surrounded by experts.

Rosenberg argues that it's the same with artificial intelligence. If you concede your knowledge (and your power) when you shouldn't, the group gets less smart. The question he raised, and it's a powerful one, is: What happens when we treat AI with the same kind of skepticism that we would give to anyone else?

Human beings tend to believe that AI is perfect. It's not. Like humanity, AI is evolving. Getting better. But that evolution doesn't exist in isolation. Without a prompt, AI doesn't have much to offer. At least, not yet.

Humans can live without AI, Evan reminded himself, but AI cannot live without the human. The human, he realized, is what feeds and guides the AI. His job was to set up the people and process to guide outcomes. The organization is really a blended sort of swarm where people were optimized, so that output would increase.

He wanted a human-centric organization, where AI accelerated results. He wanted a design that was sustainable - not churn-and-burn, where people are cut out of the organization with a machete. What he wanted was Human Success. And these scores and these evaluations that Rasmus' team was sharing with him: that was the pathway to a new number. A new way of guiding and leading and interacting - Evan sensed it, deep inside. Maybe quitting wasn't his only alternative here.

He pulled out his phone and excused himself for a moment, stepping to the railing at the edge of the large patio. From there, he could easily look down at the ebb and flow of the Hudson River, 24 stories below. Using an API that linked messages to a chatbot, he recorded a quick message to his assistant, Gabriela. He told her where he was, the patio on the 24th floor, and let her know that he was not to be disturbed unless something was really important. He asked her to keep an eye on emails as he would be occupied at the conference for a few more hours. Then he sent a message to his team. The senior executives that were traveling with him in New York needed to meet him in the lobby before Robin's keynote. He had something he wanted to discuss before the kickoff event.

Back in Texas, Lachlan Moore was ready to give in to AI in exchange for profit. In the short or long term, Evan wasn't sure that trade-off was going to work. Rosenberg had cautioned against seeing AI as equal or superior to humans, in his writing and in interviews. Evan felt Lachlan Moore believed that AI was interchangeable with human beings. AI could be considered equal (or superior) factors of production - but Evan didn't see the equation in the same way. At the very least, Lachlan seemed willing to play fast and loose with what AI could do. Evan wanted to measure twice and cut once.

Because the algorithm was working, 24 x 7, and never needed to take a break, Lachlan saw the relentless speed and output. Was he able to see the workshop as well, Evan wondered?

Lachlan's logic: Manage the mistakes, embrace the always-on human substitute, and capture massive rewards. Cut the people, replace with profit.

Rosenberg had a different view. "It's dangerous to be replacing critical decisions," he said in an interview, "decisions that affect organizations and ultimately, society, with AI. And my big push is to keep humans in the loop."

Evan felt the same way. It wasn't some Pollyanna ideal. It was just good business.

He headed back to the table to rejoin Rasmus and his colleagues. As he moved to sit down, his phone rang.

He sat down at a nearby table, away from the group, to take Gabriela's call. As always, her timing was impeccable. She wanted Evan to know what was coming. And what she had been asked to do.

Taking six steps back to his chair, and the group, he saw a distinguished gent in a sport coat walking towards him. In his hands was a large blue box, with a thick white ribbon wrapped around it. As he approached, Evan could see that the blazer was his uniform, and he was an employee of the hotel. His nametag came into view. It said, "Max" in black letters, on a gold background.

Standing next to Rasmus, Max said, "Are you Evan Sato?"

"What's this?" Rasmus asked, surprised. Nina leaned in.

"Uh, that's me," said Evan, raising a hand to Max.

"This is for you," he said, placing the box on the low-slung table where Evan was seated.

Max quickly came around the table, reached into his breast pocket, and pulled out a small envelope. He bent down and spoke into Evan's ear. "This is for you as well," he whispered, placing the card in Evan's hand.

He opened the thick cotton envelope and pulled out a note.

At that instant, Robin returned. He had finished his sound check next door.

Approaching the table, he saw the blue box. It was unopened but already oozing luxury by its crisp appearance. "Whoa, look at this! What are we celebrating?"

Good question, Evan thought to himself. Did he just receive a parting gift? Would the note say, "Thank you for your service. And your pushback in the board call. *Adios*." Or what exactly?

Evan opened the tiny envelope and read the contents with calm curiosity. The note was from none other than Lachlan Moore. That meant the gift was as well. Was this another ambush, he wondered? The heavy cotton card stock was a customized letterhead from his private firm, Great Scot LLC. Evidently, Lachlan was in New York. And he wanted to talk with Evan face to face.

The note asked for a meeting at the hotel bar in just a few hours, after Robin's keynote.

Evan was already thinking about what he might do or say in such a meeting when Gabriela called him. In their recent phone conversation, Gabriela told Evan that Lachlan had reached out to her. He called her on her cell. The Scotsman peppered her with curiosities about Evan's whereabouts. Something was up, she told Evan. She wasn't sure why Lachlan was asking so many questions.

· · ·

E van looked at the card in his hand. The note said the box contained a peace offering. Evan wondered if the conversation at the bar would be a peace offering? Perhaps Lachlan wanted to bury the hatchet. In Evan's back.

Evan slid the note back into the envelope and put it in the vest pocket of his sport coat. Whatever the case may be, Lachlan was nearby. Probably at the Equinox Hotel.

Lachlan was sending a message: "I see you, Evan. I'm watching."

Looking at the box, Evan realized something: Lachlan was good at saying he was sorry. He suspected the Scotsman had had a lot of practice. After all, you can't get to be in a position like Lachlan's without understanding how to work with people – especially if your personality could be a liability at times. Apologies and resets were business as usual for Lachlan. The box and the card: Evan saw the whole thing as an invitation – not an apology.

He untied the ribbon and opened the box, sliding the cardboard cover away. Instantly, the deep rich smell of 100% British bridle leather filled the air. Robin got out of his chair and came around the table to get a closer look.

A natural cotton dust bag hid something substantial – something in a rectangle shape, about 20 inches across and 16 inches wide – that he could smell but could not see. Evan removed the bag, slowly, revealing a Barra black leather laptop holder from McRostie of Scotland. His initials were on the outside, embossed in belt that passed effortlessly under the single silver buckle on the front. The portfolio was sturdy, minimalist, and stunningly beautiful. It was perfectly sized to hold a small laptop and a good number of documents. "Designed for people on the move," Evan read, out loud, as he looked at the accompanying marketing materials.

Rasmus and Nina oooh'd and aaaah'd over the gift. "Nice," Rasmus said. "I'd like one of those!" Nina joked.

Evan read it again – "designed for people on the move". That sounds like me, he thought to himself. "It seems," he said to his friends, "that I've got a date with an investor, after Robin's keynote."

**T**he Personalized Playbook

1. **Measuring the Immeasurable:** The Human Success Score (HSS) links people-sentiment and capability to financial outcomes. What system or systems do you have in place, right now, to allow your organization to measure progress in this fashion? What is one critical "soft" metric (like team trust, decision-making speed, or innovation culture) that your organization currently dismisses, but which, if quantified, would have the greatest positive impact on your P&L statement?

2. **Manager Impact Score:** The Human Success framework aims to make manager impact measurable, moving beyond just team outputs. How are your current middle managers measured? Are they rewarded for efficiency and control, or for enabling and augmenting their team's skills for the age of AI? What three questions would you add to a manager's performance review to track their *impact* on human success?

3. **The Peace Offering Test:** Lachlan's "peace offering" is almost certainly a strategic maneuver. As a leader, what is your approach when a known antagonist offers an apparent truce? Do you assume good faith, or do you view the interaction as a strategic opportunity to gather intelligence and set the terms of the next engagement? What do you predict Evan will do?

4. **The Courage of the Unknown:** What is the biggest, high-stakes conversation your organization is currently avoiding (with a competitor, an investor, or an internal team) because the outcome is uncertain? What single piece of data (the HSS, for example) would give you the confidence to turn and engage?

CHAPTER 7

INTO THE SHED

"The killer app for AI is humans," Robin Daniels is saying. He crosses a 40-foot proscenium stage, standing in front of an enormous screen that covers the backdrop behind him. From the McCourt Theater at the Shed in Manhattan, a thousand people sit and listen to his words. Daniels is athletic and deliberate in his movements across the stage. He wears a well-fitted sport coat, jeans, a dark t-shirt and the brightest pair of running shoes you have ever seen.

"It's not about AI taking over—it's about AI making us superhuman," he continues.

Seated in the audience, about 15 rows back, Evan is wondering about timing. Not Robin's timing - it's pretty good. He's thinking about the timing on how to really embrace AI. And how to do it right. When, exactly, does AI make us super-human? And how do we embrace it exactly? Does "embrace" lead to "replace", he wondered?

The Shed at Hudson Yards.

To Evan's right, his team of six employees - including his marketing manager and territory sales leaders - were looking up at the stage.

Before the keynote started, Evan had met with his team in the lobby with a simple message. The message focused on change.

He told them that change can take two forms. One type of change can be uncertain, and scary. People fear that kind of change. "You know how we just came up a series of escalators," he said, referencing the interior design of the Shed, which relied on escalators that seemed to lead to nowhere - until you reached the top. "We couldn't necessarily see the next destination, or the next floor. We couldn't see this lobby. But we kept going. And here we are.

"Many folks are looking at AI, and digital transformation, from that same context. The change brought about by AI is looking into the unknown. However, there's another way of looking at change: change can also be progress. AI can carry us forward."

Evan went on to say that change is always filled with uncertainty (because that's how any kind of change works). From a science experiment to a conversation with your partner, spouse or supervisor - you never know how it's going to turn out.

"While no one can predict the future," Evan told his team, as they sat in chairs in the lobby area outside the theater, "we can influence it. We have the power to shape it. To shape how we integrate AI. Are you willing to step onto that escalator with me? Because I need you. Your teams are going to need you. And here's why: because we are going to step into something called Human Success."

He looked at his team, as they nodded in response. Like that cascading escalator ride inside the Shed, Evan knew that they had to keep moving. AI was transforming business, and its capabilities tomorrow would be different than today. Every time he spoke about AI, he asked himself the same questions: What didn't he know? What wasn't he seeing? And the biggie: How much could he really tell them about what's ahead? Human Success was what would bring them into the unseen future.

Evan remembered the words of his father, from a time when they were working together in the woodshed. *"Ichi-go ichi-e,"* Hiroshi said. The Japanese phrase meant, "One time, one meeting". It was a saying to help remember to treasure every moment, every encounter, as a unique and unrepeatable experience.

Hiroshi handed Evan a small chisel. *"Here. This moment is what we have, Yee-un. Nothing more, nothing less. Kono michi ya, itsu ka kita michi,"* he shared. Translated, the phrase meant, this road will someday be the road you have come from. *"Take responsibility for the choices you make now, my son. The present shapes the future."*

Young Evan looked at the chisel in his hand. He hesitated. He looked at his father, unsure where to tap into the block of wood. His father passed him a mallet shaped hammer. *"Uncertainty is everywhere. Use what is known to get to what is new. Use the tool. Touch the wood. What needs to be done, so that you can make what you see in your mind? Take the first step. Nothing more, nothing less. And the next will reveal itself to you."*

From Evan's fear of the unknown came a deeper understanding. The understanding that he could face the unknown and always find a way through it. That's how resilience works, he reminded himself. One step at a time.

"Listen carefully to what Robin has to say," Evan said to his team. "He's going to share insights into Human Success - a new philosophy for leading and collaborating in the age of AI. Listen for what you might discover, inside his words. Listen to his examples. Then personalize and apply what is said. Consider how we will apply these ideas, as they resonate with you." He didn't want to say more.

Inside the theater, the sound system carried Robin's words

to every corner of the theater. As he looked down the aisle at his team, he wondered what words would land for them.

"AI is an enabler, not a replacement," Robin was saying from the stage.

Evan wondered: yes… but for how long? Lachlan saw AI as an opportunity for subtraction. Robin was saying it was more suited to multiplication.

"AI can handle hard skills—like data synthesis, pattern recognition, and automation—soft skills like empathy, communication, and leadership remain uniquely human. This balance is central to the Human Success model, where technology supports, but doesn't overshadow, the human element," Robin continued. "Here's where AI can help."

Robin Daniels onstage.

On the slide behind him, he outlined a step-by-step process for implementing Human Success.

- AI-guided coaching for managers

- AI-enabled career pathways for employees
- AI-powered organizational insights for C-level executives

These ideas were offered as examples of how technology can simplify workflows and enhance learning. "These innovations aren't just about efficiency—they're about creating better leaders and more cohesive teams," Robin said.

Evan wondered how they were going to do that. How could AI be seen as a partner in success, not a threat? Especially if artificial intelligence is perceived as the reason for layoffs?

The slide on stage changed, to display the headline: "Soft Skills as the Differentiator". Robin moved towards center stage as he spoke.

"In a world where AI is disrupting hard skills at lightning speed, we need to double down on soft skills. The way you communicate, collaborate, and bring people together will be the superpowers that define great leaders and teams in the future. AI can't replicate the emotional intelligence required to lead with empathy or have tough conversations with a team."

Indeed, Evan knew this all too well. AI could create a slide deck, but it couldn't deliver the presentation for you. AI couldn't respond to Q&A, in the room, in real time. Because AI struggled with context. Empathy. Ethics.

People skills were, for the most part, still the domain of people.

Without the ability to see, hear, feel and empathize with

what is happening in real time, *ichi-go ichi-e* eludes AI. Because it can't read the room. AI does not have a moral compass, an ethical code, or a viable sense of consequences. The LLM wasn't present in the moment. It was learning and pulling from the past.

"AI can significantly enhance decision-making by providing faster access to insights and data," Robin shared. "In many cases, it's so fast and has access to so much information that it can appear like it knows more than anyone on earth. It probably does!" he said, as a ripple of nervous laughter emerged from the audience. "But the how and why behind decisions—especially those involving people—still requires a human touch."

"AI can guide, but humans must lead," Robin said.

Evan looked at his team. As Robin spoke, a new strategy began to emerge for him. The CEO, the one who wanted to quit, was finding another way. He sat up straighter in his seat.

"AI isn't here to replace us—it's here to make us better. Smarter. Faster. More human. And if we get this right, we're not just building better organizations—we're building a better world." Robin was smiling. He stepped down stage, closer to the audience, and held out his hands, palms up.

"A team is a collection of people trying to achieve great things together. Team alchemy is the goal," he continued.

"As leaders, we want to do something epic," Robin said, nailing one of his favorite words with a conviction that was… well, epic.

Evan's mind drifted. He thought about telling Lachlan that he was done. That would be epic, wouldn't it?

Epically stupid, he reasoned.

The six people on his right were just a fraction of the folks who were counting on him.

"Here's how to be truly epic," Robin said, simply. "Either find something that's broken and do it so much better. Or do something that hasn't been done before."

At that moment, Evan decided he would do both.

————

The Personalized Playbook

1. **Is AI a formula for Subtraction or Multiplication?** Which of these two views dominates the strategic narrative in your leadership team? What single, tangible initiative could you launch this week that explicitly uses AI to multiply human effectiveness rather than subtract costs?

2. **The New Epic:** Robin states, "Either find something that's broken and do it so much better. Or do something that hasn't been done before." Which path is less risky for your organization right now? If you chose the latter (something that hasn't been done before), what is the first, small "proof of concept" that would demonstrate the power of your Human Success approach to your Board

3. **Beyond the Black Box:** Robin emphasizes that the "how and why behind decisions" still requires a human touch. In your current AI adoption efforts, what specific leadership decisions (e.g., hiring, budget allocation, culture shifts) are you allowing AI to guide, but ensuring decisions are ultimately led and owned by a human who can explain the *why*?

4. **The Escalator to the Unknown:** Evan used the Shed's escalators to describe the uncertainty of the future, and its unpredictability. As a leader, how clearly have you communicated the destination (the vision of the AI-augmented future) to your team? What is your strategy for ensuring they focus on the top floor rather than the anxiety of the ascent?

5. **Team Alchemy:** Robin's ultimate goal is "team alchemy"—achieving great things together. What one cultural or operational barrier is currently preventing true alchemy (the synergistic combination of talent) within your organization? How would the Human Success Score help to identify and remove that barrier?

PART TWO
BATTLE LINES

"**A**m I out of time already?" Evan asked.

He stood right in front of Lachlan Moore's table, unnoticed. The restaurant at the Equinox Hotel was still doing a brisk business, as diners and drinkers were seated in wooden Danish dining chairs, sharing conversations across dozens of white-tablecloth tables. The lights of the Manhattan skyline flickered outside the dark windows of the restaurant.

At the table, Lachlan Moore was seated behind a tumbler of whisky, wearing an Isaia windowpane blazer. His arms were crossed in front of him, on the table. The Scotsman was looking down at his wristwatch, twisting the large dial, as if trying to memorize its features.

Approaching the table, Evan could see the timepiece from six feet away. The Panerai watch was popular among investors of a certain age. On Lachlan's thick wrist, it looked like the Bat Signal.

"What? No, no, no," Lachlan spoke as he looked up at

Evan. "New watch," he said, turning his wrist in the air to display the black and silver masterpiece on his forearm.

"Your timing is perfect," he said, coming around the table. He shook hands with Evan and made a sweeping gesture with his left hand. "Please, join me," Lachlan said, with a laugh that was apropos of nothing. "So, how was tha' keynote?"

Evan hated small talk. Not many people knew that about him, because he was so good at it. Chit chat was part of the game, so he played along. "Not bad," he said. "I learned a few things. And thank you for this, by the way," he finished, holding up the new satchel before he placed it on the seat beside him.

"McCrostie leather is th' best. Enjoy," Lachlan said.

Sitting down at the table, Evan glanced at the menu in front of him. He wanted an appetizer. But he needed information. He needed to find out more about how Lachlan snuck onto the Board without his knowledge. Glancing down at the side dishes, he expressed his surprise at hearing Lachlan on the board call.

In response, Lachlan explained that he had been watching the company for quite some time. His presence on the board was deliberate and calculated. Calculated: like Lachlan himself, Evan reasoned.

Lachlan shared what drew him to the company. Evan responded with some common ground around his own decision to come aboard, which surprised both men.

Evan and Lachlan were not completely different, even though they saw a major business issue from wildly different viewpoints. As the men talked, Evan noticed an ease to the conversation.

When Lachlan mentioned that he saw Evan as a one-of-a-kind CEO, Evan was surprised. "I'm very curious to hear about your discoveries," Lachlan finished. "And what you're gonna do next."

Me too, thought Evan.

What he didn't know was that Lachlan wasn't trying to compliment him. The Scotsman favored facts and strongarm tactics over emotional ploys like flattery. Lachlan was honest - often brutally so. He inserted himself on the board, he explained to Evan, because it was time for the company to focus on EBITDA, not just top-line growth. The shift, in Lachlan's mind, was a logical progression.

Lachlan's Board position, in Evan's mind, was an ambush. Even though, Evan knew, the investor's viewpoint wasn't totally wrong.

Evan accepted the Scotsman's board presence as a given. And, given that Lachlan was on the board, the two men needed to work together.

In the board meeting, Lachlan had asked Evan for his "number". A number that would create shareholder value, minimize costs, and maximize output. Lachlan's number was 40% - the size of the workforce cuts he demanded.

"My father was a woodworker," Evan explained. "He owned a manufacturing business that made curios and souvenirs out of wood and other natural materials. But his passion was woodworking.

"I'll never forget the time when he told me about the most important tool in his workshop," Evan shared, recounting the story of how the real power rests in the carpenter's hands. "But then he surprised me. Because he also asked me, 'what was the second most important tool?'"

"Th' second most important tool?" Lachlan wondered.

"Yep. By this point I thought it was a trick question, so I said something like 'the button that turns on the machine' or something like that. But my dad shook his head. He reached into his back pocket and handed it to me." Evan leaned in for emphasis. "You have to know: my dad was always talking to me in Japanese. That was actually my first language. So he goes, '*Makijaku.*' And points at it.

"*Makijaku* means tape measure," Evan said. "*Makijaku de hakaru,*" Evan continued, looking out at the windows but seeing the memory of his old man. "Measure with the tape measure. That's what he would always say, over and over again."

"If you don't measure, it doesn't matter how beautiful your wood may be, or how brilliant your cuts and grooves are. You can't just eyeball it and hope for the best." Evan looked away from the Scotsman and stayed with the memory. "Take a curved staircase, for example, where you are creating a wood banister. We would do projects together like this, when I was just starting high school. Without measurement, pieces don't fit together. You've got to think about vertical rise and rotation – so there's a precise mathematical slope. Calculations and measurement are the only way to do it. If you're off by even

1/8 of an inch," Evan explained, "your handrail can be as much as two inches too high when you are trying to meet the floor joists on the second story.

Remembering an old project… with a new twist.

"We did this together for a friend who owned an old house where the bannister had become worn out and unsafe. I was about 15, I guess. If we were off by the slightest amount, we could create a 'joinery nightmare' where things could not connect," Evan stated.

"I don't want to create a disjointed nightmare at this company. I don't want to cut without measuring. Otherwise, we lose our ability to connect. A handrail that doesn't fit the pitch of the stairs becomes a safety hazard. A carpenter's nightmare. And probably a lawsuit, if it's in a residence or commercial building. Moving quickly can keep us from doing what we are supposed to do. In fact," Evan said, folding the menu, "I won't cut without measurement."

Before Lachlan could interject, a waiter appeared out of nowhere. "Drinks, gentlemen?" They ordered, and as the waiter walked away, Lachlan said, "Tha' reminds me of a little Beer."

"Did you say, 'beer'?" Evan wondered. That's odd, he thought.

"Ay, I'm talkin about Stafford Beer, the English social scientist. I hate t' give credit to the English, but Beer said something that kinda sticks with ya, ya know?"

Evan thought to himself, no I do not. Who is this "Beer" guy?

"Beer said, 'The purpose of a system is what it does, not what it claims to do'," Lachlan paused to see if the quote had landed. "It's about lookin' at outcomes, not intentions," Lachlan said.

Evan nodded. "I like it," he said. "So if a system isn't working as intended, it's still working towards its real purpose. Even if that outcome isn't what you want, it's what the system is designed to do."

Lachlan nodded. "Exactly."

"So are you trying to say that we need to change the structure of our organization," Evan wondered, "instead of blaming its participants?"

Lachlan raised an eyebrow, as the drinks arrived on the table. "What does your tape measure tell ya?"

As Rasmus and Evan were walking out of Robin's keynote, just minutes before, Rasmus had turned to Evan with a similar question.

"One thing I'd like to know," Rasmus had said, when they were walking from the Shed to the hotel, "is how much time you have to fix this thing."

"Yeah," Evan said, nodding. He asked his friend, "How much time would you ask for?"

The two men left the lobby of the Shed and crossed the street towards the hotel.

"I've got a lot more to tell you," Rasmus said, "about how Human Success works. But for now, if I were in your shoes, I would ask for one year."

Evan shook his head. "I don't think they will go for it."

"Then find out what they will accept," Rasmus said. They were almost at the door to the hotel. "Once you get a timeline, you have to decide if you can do it," Rasmus said.

Or if I want to, Evan thought to himself.

"There's something called the Productivity Paradox," Evan said to Lachlan, recalling what Robin had shared in his keynote. "Technology increases efficiency, but it doesn't always translate to better outcomes."

Lachlan nodded. He wasn't thrilled with Evan's approach, as Lachlan was usually the one doing the lecturing. But he didn't really know Evan, and wanted to give him some space to maneuver. Plus, Evan's style wasn't really a lecture - it was more of a sharing between experts. And an attempt to bring new ideas forward.

"There are several studies that show that AI assistance does not necessarily translate into greater output. In fact, one study conducted by the Model Evaluation and Threat Research Institute showed that using AI tools actually decreased efficiency by 19%" Evan shared. He was careful not to instruct his newest board member, but he needed to get a fresh context into the conversation.

"I know, I know," Evan continued, raising his hands. He was introducing potential objections before Lachlan could get to them. "This survey was in a different industry, the study was an isolated incident, your mileage may vary," he gave a half-chuckle. He wasn't going to deny the impact of artificial intelligence - or debate the study. He just wanted to explain that there's a right and a wrong way to embrace AI. It was a tool, not a deity. Evan wanted to explain how he saw the difference between the hype and the reality of working with AI.

"Many people are reporting positive results with AI," Evan said, lowering his hands. He went on, "There's a point that is universal here, and that's what we need to discuss. In this same survey, developers *believed* they were more productive with AI. But they weren't. The takeaway is that there's a difference between perception and reality when it comes to working with LLMs," Evan said. "In fact, the MIT Media Lab reports that 95% of organizations see no measurable return on investment on these AI technologies."

"*Yet*," Lachlan interrupted.

Evan nodded. *Touché*. The point was well taken. Inside that single word, Lachlan spoke volumes.

"Google's DORA report says that AI can act as a 'mirror

and a multiplier.' In cohesive organizations, AI boosts efficiency. In fragmented ones, it highlights weaknesses."

"Hmm. Which one are you runnin'?" Lachlan asked.

Evan leaned forward in his chair. "Lachlan, I share your goals. Your ideas are solid. Well thought-out." Evan spoke with conviction in his voice. "How we get to the goal matters to me. I don't want to run a bunch of digital experiments, or cut people without consideration, as these approaches are going to lead to regret, not profitability."

"Ah. The 'let 10,000 flowers bloom' approach," Lachlan said, referring to the business strategy where many expensive experiments were needed to produce a small handful of highly profitable returns. The investor took a sip of the Lagavulin 25 that was sitting in front of him.

"So, AI is a double-edged sword," Lachlan said. "Especially when it comes to people. So I've heard." Raising his glass, Lachlan toasted Evan with his tumbler.

Lachlan wondered when Evan was going to say what was really on his mind.

"I don't really do context blindness," Evan said, the volume in his voice rising ever so slightly, "where I make bold moves without considering the culture of the company. Or the repercussions."

"Maybe you should start," Lachlan said. A wry smile - a *gallus* smile - parted his lips. "Fortune favors tha' bold."

Evan responded with a smile of his own. He leaned back in

his chair. "Boldness," Evan said, nodding. "That's something we can definitely agree on."

Perhaps good Scotch was another? Evan swirled the glass of Oban 14 in front of him. Staring at the product of human craftsmanship in his glass, Evan found himself balancing tradition with innovation. He held both, in his hands and in his head.

"I want to make some big changes at the company," he said, hoping that Lachlan wouldn't dive in on the specifics. He leaned into the table, tasted his scotch, and continued. "Your focus on EBITDA, operations - all of that - I'm on board 100%. But these changes take time. I've spoken briefly with my team, but my real plans have only just begun to take shape. We need to do this together."

Evan was watching Lachlan for a flicker of impatience, or a hint of disappointment. The operative word here was "we". Was Lachlan an adversary, or an advocate? Was he going to give Evan the time he needed?

"Go on," Lachlan said.

"I want to bring you more than just a number. I want to share our playbook - a playbook for the entire organization. That way, we build this thing the right way and we create a sustainable company."

Lachlan nodded. "A playbook is a good start. But it's not the same as runnin' the game. Th' map is not the territory," he said. "Action matters. And results matter most."

"I am going to need time to put this plan in place and restructure the organization," Evan finished.

At last, thought Lachlan. There it is.

"I unnerstand," Lachlan said. He signaled the waiter for the check. "You've got six months."

———

T he **Personalized Playbook**

1. **The Operative Word:** Evan consciously uses the word "we" ("We need to do this together", e.g.) to invite his adversary into a partnership. When facing internal or external opposition to a change initiative, what is the single operative word or phrase you can introduce to pivot the conversation from "you vs. me" to "us vs. the problem"?

2. **Negotiating Time:** When facing a short-term financial demand (like quarterly targets or immediate cost cuts), how do you calculate the minimum viable time needed to implement a sustainable, Human Success-focused solution that satisfies the short-term demand without destroying the long-term culture? Could greater insight into the Human Success Score help with your ability to deliver, and hit projections?

3. **The Language of the Enemy:** Evan found common ground by acknowledging Lachlan's focus on EBITDA and operations. What is the single most important financial or operational metric that your company's most demanding stakeholders (investors, board, competitors) prioritize? How can you use the

Human Success framework to show that investing in your people is the most efficient and sustainable way to *improve* that metric?

4. **Beyond the Golden Rule:** Lachlan represents the traditional power structure. Evan is asking for time to replace that structure with a Playbook—a governance model based on data and trust. What steps must you take to ensure your "Playbook" is viewed by stakeholders as a serious, auditable, and financially rigorous governance document, and not just a soft set of HR initiatives?

5. **Frenemy Management:** Evan understands that he must transform his adversary into an ally, or at least a neutral observer. Identify a "Frenemy" (a skeptical leader, a competitor, a challenging investor) in your ecosystem. What shared, aspirational goal can you establish with them that requires them to grant you the time and support needed to implement your own "Human Success Playbook"?

CHAPTER 9
THE EXECUTION CRISIS

The next day, Rasmus and Robin were walking towards their hotel after an early morning run. The sun was rising in between the skyscrapers, throwing long shadows across the pavement, as they crossed near the subway stop for the 7 train. Their hotel featured a 60,000-foot gym, but the men wanted to get some fresh air. Turns out, they were not alone in that regard.

They walked across a large concrete plaza that was already filled with people headed to work. To their left, parked on the plaza pavement, a food truck was doing a brisk breakfast business.

Standing beside the silver truck was none other than Evan Sato. He had just gotten a coffee and turned around when he saw Robin and Rasmus.

"Gentlemen! Good morning," he said. He wore jeans and a lightweight blue hoodie. His voice was strong, and his smile was wide.

Rasmus picked up on it right away. "You're in a good mood!" he said. "What's going on?"

Robin jumped in, "How did your meeting go last night?"

"Overall? I'd say it was terrible," Evan said.

"What? Then why are you smiling?" Rasmus asked.

"Because I know where I stand," Evan said.

For the first time, Evan saw how the pieces would fit together. Like a scientist watching an experiment, he listened to Lachlan with a detachment that had served him well in his career. He had tried to set the stage for a new understanding with the board. He wanted to talk about Human Success, but Lachlan was unwilling or unable to listen. Evan realized he didn't need to teach Lachlan how he viewed AI. Like an expert carpenter crafting a handrail on a custom staircase, he didn't need to explain the process of cutting and shaping to create the final product. He needed to implement the process, so that the pieces all fit together in a way that was useful. Then, and only then, would he accomplish the goal. Implementation made the transformation complete.

He knew how to do that. And he also knew that, if he didn't, he would probably be one of the first people to go.

Evan knew he could be pushed out. But he wasn't going to jump ship.

"I know what I need to do. And I'm going to need your help to get there."

Rasmus turned to the small counter on the food truck to grab a couple of water bottles, tapping his phone to pay.

Evan explained to Robin that Lachlan had given him six months to sort things out, as a kind of ultimatum. "What did you say?" Robin wanted to know.

"Well, I tried to offer some points of reference as to how AI might not be the panacea, or a replacement, but I'm not sure it landed at all. He threw out the 'six month' number. I pushed back, just a little bit, and got him to agree to wait till the AGM - the shareholder's meeting. So, it's roughly 8 months from today - in Long Beach.

"I need to show a new plan. New milestones. The whole nine yards. I told him that it's not about headcount. I can get rid of headcount. But in the world of AI, I need to double down on certain kinds of people, like you were saying."

Robin smiled at that remark, because it was one of the key themes in his keynote. "But the question is," Robin said, continuing the thought, "which ones?"

"I told Lachlan we might need to cut more than 40%," Evan said.

"What?" Rasmus wasn't expecting that.

Evan replied, "I was just saying that we need to get under the hood before we pick a number. I'm here to do whatever is necessary. We need to know who and what we have, before we start slicing into the organization. Before turning to AI for answers that it doesn't yet possess."

Rasmus passed a water bottle to Robin. "Look at this," he

said. The men moved around a wooden bench, to sit at a sea foam green table near the breakfast truck.

"I got this email yesterday from a friend of mine. She's the Chief Revenue Officer at a tech startup. Check it out," he said.

He pulled his iPhone from the pocket in his Tracksmith running shorts. Robin and Evan moved even closer to the text - and they didn't like what they saw.

Hey ██████████ *– I wanted to share some **direct feedback** ahead of our Friday call, because I care about your success and want to help you grow into the **strategic leader** I know you can be.*

Ramus interjected, "The leader who is supposedly sending this message is from eastern Europe. This boss does not talk or write like this at all. My friend was immediately suspicious."

"Look at all those bold letters. Hmmm…smells like AI," Robin said, drifting closer to the screen.

*Right now, I'm concerned about your level of **proactivity and ownership** in the role. To be blunt: I feel like I still have to do too much of the thinking, identifying problems, and driving change — and that's not how it should be at your level.*

For example:

*We don't support **CloudFormation drift**. That's something you could have just delegated to Karthik or Juana to update the docs and align with SEs, without looping me in. Instead of having long discussions about the roadmap, you could go into **BlackMarble**, generate one, and ask Marcella if it's good enough.*

"That's their proprietary internal software," Rasmus said.

"You'll see other stuff that refers to the company processes. Just notice the tone of the message," he finished. On the screen, here's what they saw:

- *When you see we're losing smaller deals on price, I'd expect you to proactively suggest, "Let's test **X and Y** for this segment."*
- *With our big bets, I think you could be playing a much stronger **C-level role** — owning progress, bringing energy and alignment — rather than escalating things to me.*

*More broadly, I've been the one digging into data to uncover gaps: that we're not doing **value selling**, that we might not be focused on the **Ideal Customer Profile (ICP) that converts**, or that **pipeline quality** is off. I suggested **Pipeline Day**, and I pulled in key influencers for the big bet reviews. I'm the one calling out the need to look at **sales cycle data** in detail.*

All of this should be coming from you.

*We're a startup – I need you to be the one **looking ahead, spotting gaps, bringing hypotheses**, and **pushing strategic changes**. I know you're capable of this. Maybe it's a matter of reprioritization or dropping some lower-leverage tasks. But we need to talk about it.*

*Let's dig into this on Friday – I want to hear your take. My goal is to help you get to a place where **you're the one driving these conversations**, and I'm a **thought partner**, not the initiator.*

Evan looked at Rasmus. "What did your friend do when she got this email?"

"She went directly into her boss's office," Rasmus said,

"and told him that if this is what he thinks of her performance, she's at the wrong company."

"What?" Robin said. Evan blurted out, "Wha - She quit?"

"She quit. On the spot," Rasmus said.

He went on, saying, "When we have people turning to AI to write and summarize evaluations, it doesn't help anyone or anything. If we are all passing AI documents around and getting AI responses, we are not really moving."

Robin nodded his head. "Because we are not connecting."

Rasmus wasn't done. "In fact, her boss didn't even know what the AI had written in the email - he was shocked. Because when she showed it to him, he was reading it for the first time. He tried to walk it back. He tried to explain that the AI was just responding to prompts and summarizing meeting notes. But the damage was done - and so was my friend. She didn't care if an AI wrote it and his assistant sent it. No one could un-ring the bell."

Evan reflected on the message, and its implications. AI-generated reviews. AI-generated reports. AI-generated résumés. The review of all those things, and more: done by AI. Things were easier, but were they getting better? Was AI a connector, or a divider? Was it a service, or a shield - keeping us from seeing what really mattered?

In that email, the generic tone was tone deaf. Evan couldn't afford that kind of exchange inside his company. He couldn't afford to turn certain aspects of his business over to AI. At least, not entirely. Where AI added speed and efficiency, he'd be a fool not to use it. But managing how AI integrated within his processes was really the question. AI without humans in the equation did not add up. The AI solution was incomplete without Human Success at its core.

"So, gents, how do we implement guardrails, so that AI is used in ways that don't cause our best people to quit? Because we are already using AI throughout our organization," Evan said. "And for me, the clock is ticking. I've got less than eight months to get competitive with AI-centric companies."

Evan needed a strategic plan to counter-balance Lachlan's attachment to slashing headcount. The plan had to focus on wise decisions around AI, with people at the center of the conversation. But having a plan isn't the whole story.

Once again, Evan said the quiet part out loud. "The real question is: how do we execute on the plan?"

At that moment, Nina walked up to the table.

"Who's getting executed?" she said with a smile.

Nina's put-together appearance was quite the contrast to her two sweaty co-workers. She wore a bright blue jacket and a thin scarf around her neck. She had a large leather satchel over her shoulder. In her left hand, a cup of coffee.

Rasmus and Robin smiled. Nina smiled back at her co-workers, and at Evan. "Your timing," Robin said, "is perfect."

"I couldn't help overhearing your conversation," Nina said, looking at Evan as she raised her coffee cup. "What you need is The Trifecta."

. . .

Evan wasn't sure if The Trifecta was the name Nina's coffee drink, a big bet at Belmont, or what exactly. Nina's confident tone made it sound like she thought it was a sure thing, whatever it was.

Evan pulled an empty chair around and gestured to Nina to take a seat. He said, "I've seen so many brilliant strategies fail because the culture refused to adopt them."

Rasmus leaned in. He put his hand on his friend's shoulder. "Evan. You're going to like this part," he said.

———

The Personalized Playbook

1. **Clarity Over Comfort:** Evan smiles because he "knows where he stands" after the terrible meeting. In your organization, where are you prioritizing comfort (avoiding confrontation, seeking consensus) over clarity (defining the opposition, setting non-negotiable goals)? What difficult, clarity-inducing conversation do you need to have this week?
2. **The Execution Crisis:** Evan's crisis is: "how do we execute on the plan?" Identify a brilliant strategy your organization developed in the last year that failed to achieve its full potential. Was the failure due to a flaw in the *plan* or a weakness in *execution* (e.g., lack of manager training, poor communication, or cultural resistance)? What do you make of the famous quotation, "Man plans, God laughs"? Explore how adaptability fits into your plans.

3. **The Trifecta of Action:** Nina promises a "Trifecta" to solve the execution crisis. If you had to identify the three non-negotiable pillars (e.g., executive alignment, middle manager empowerment, real-time data measurement) that your organization needs to successfully implement a major culture shift, what would they be?

4. **The Clock is Ticking:** Evan turned six months into an eight-month deadline. If you were given a similar window to completely transform your organization's approach to AI and human capital, what is the single "quick win" (a measurable, high-impact project that takes less than 60 days) that you would use to build trust and momentum for the larger Human Success Playbook?

5. **Daiku and Fault Lines:** Evan's clarity stems from understanding Lachlan's fault lines (binary thinking). What are the fault lines (deep-seated beliefs, political rivalries, or historical failures) within your own organization that will actively resist the execution of a new, human-centered approach to AI? How can your new "Playbook" address these fault lines directly?

CHAPTER 10
THE KATZENBACH CONNECTION

"**B**ut first," Nina said, "let me give you a little backstory."

Robin and Rasmus excused themselves to go get cleaned up. The conference didn't start for a few more hours. Soon, Evan would meet up with his own team and they would begin a series of customer and distributor meetings. What Evan didn't know was that, in the short time before the conference started, Nina would completely reshape his thinking around organizational design. But the transformation ahead would require him to confront a problem that was bigger than anything Lachlan had thrown at him so far.

Nina opened up her laptop and showed Evan a single image. The image was from a book by an author named Katzenbach.

Investigating Katzenbach.

"What do you know of this guy, Katzenbach? Ever heard of him?" Nina said.

"Sure," Evan said, standing up to toss his coffee cup into a nearby trash can. "He's the co-founder and Chairman of DreamWorks."

Nina laughed just a little. "You're thinking of Jeffrey *Katzenberg*. Jon Katzen*bach* was a McKinsey consultant and author, who wrote a best-selling business book called *The Wisdom of Teams*."

"Ah," Evan said, nodding. "Oh, yeah. I remember this guy." Nina was a student of organizational behavior: a wizard with the guidance Evan knew he needed.

He recalled reading chapters from the book in some organizational structure class at Berkeley. Slowly, it came back to his memory. "He was the guy who talked about common sense and uncommon sense." Nina smiled. Of course Evan was familiar with *The Wisdom of Teams.*

Evan scratched his left ear, trying to remember more. He was distracted by the common sense of financial panic but revived by the uncommon sense of focusing on Human Success. "That Katzenbach book said that teams outperform individuals acting alone, especially where performance requires multiple skills, judgments and experiences."

Evan was surprised he remembered that tidbit. But then he recalled that there were a lot of collaborative assignments at the Haas School, where he got his MBA. Talking about teams was an important part of the discussion in classrooms there.

"What exactly does a business book from the nineties have to do with structuring an organization in the age of AI? Haven't we moved on from *The Wisdom of Teams?*" Evan asked.

"Some wisdom is kinda timeless," Nina shared. "Especially when it comes to a sense of shared purpose. And human success."

Evan wasn't sure about that last remark. He tilted his head to the side. "What do you mean?"

"This pyramid formed the centerpiece of the book," Nina explained. "It's the unified and purpose-driven approach to maximize team output." Nina could tell by the way Evan looked at the screen that this information was new. Or perhaps newly remembered?

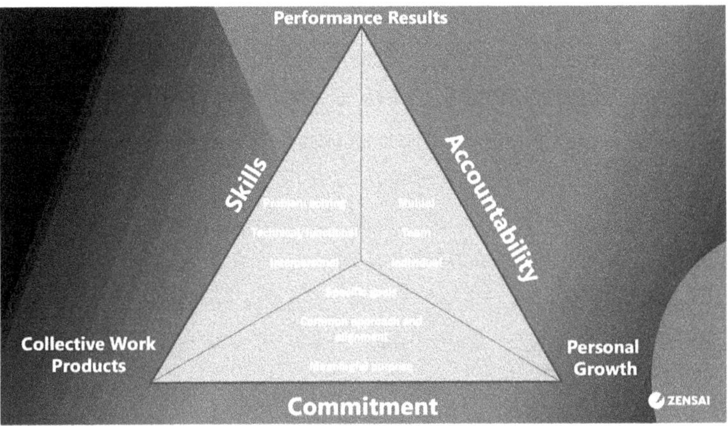

"Like what Robin said in his keynote," Evan offered, as he tried to make sense of the image. He quoted Robin verbatim: "Team alchemy is the goal. And a team is a collection of people trying to achieve great things together."

"That's exactly right," Nina said. "Katzenbach says teams that fail are teams that are made up of individuals who are more interested in asserting themselves, rather than holding one another accountable in a true and authentic way."

"Right," Evan chimed in, "with things like radical honesty and conscious confrontation," he shared, talking about two things that were prevalent and expected in his company. He had worked hard to make sure that people could communicate, even when the topic was difficult or the conversation was corrective. He had abolished annual reviews, favoring a fast feedback model like the one Robin advocated in his keynote.

The idea was that feedback was crucial to employee success. And dealing with things in real time was priceless: no one could afford to wait for a performance review to receive an unwanted surprise, or "Gotcha!" - Evan's term for the kind of supervisor "feedback" that does more harm than good.

"So, Katzenbach recognized that teams at the top can sometimes be weaker, because their shared performance goals are not as important as individual success," Nina said.

"I get it. Like the C-Suite," Evan said, "where people are more focused on their own careers. More focused on individual achievement. But every executive understands that the success of the team is their success - teamwork is the rising tide under every boat.

"Here's how I've addressed that 'individual versus team-focus' with my execs. First, I try to create a friendly and collaborative atmosphere. Look, I know, your boss isn't necessarily your friend. But leaders can't lead without respect and the willingness to listen. Relationships are vital at the top levels. My leaders aren't there to grant employee wishes, or let the inmates run the asylum, but they are required to listen - even when it's tough. And they ask the same of their teams," Evan finished.

Nina nodded her head. She could understand why Evan was in the role that he was in.

"Then you have to set expectations in a new way," he said. "My direct reports, and into the director levels, are asked to *demonstrate* behaviors. Not just talk about them. "As Gandhi said, 'Be the change you want to see,'" Evan explained.

"Don't ask someone to do something you wouldn't at least be willing to do, even if the task is outside of your skillset or expertise," he continued. "If people are working late or on the weekend, you're not spending time on your boat while that's happening. Or if you are away, you have to find ways to stay connected, engaged and supportive. Because you are always a part of the team, when you are leading it."

"That's brilliant," Nina said. "That's exactly what Katzenbach was saying in his book. You have to establish a sense of shared responsibility. A mutual sense of shared purpose and commitment. He says that we can harness sub-teams for increased accountability."

"Sub-teams?" Evan asked. He knew the term, but he was unsure of the context.

"Yeah," Nina said. "In a smaller sub-team, members work together more intensely and frequently, building the necessary trust and shared commitment for peer-to-peer accountability to flourish. The objectives of the team, as a whole, come from a mix of skills and personalities. But Katzenbach says the skills are what matter most."

"I guess dealing with personalities is what makes it feel like work," Evan joked.

Nina smiled. "One concrete example of using sub-teams is in the manufacturing process. The sub-teams are formed to tackle specific, high-priority segments of the overall performance challenge," she said, pointing to another slide on her laptop screen.

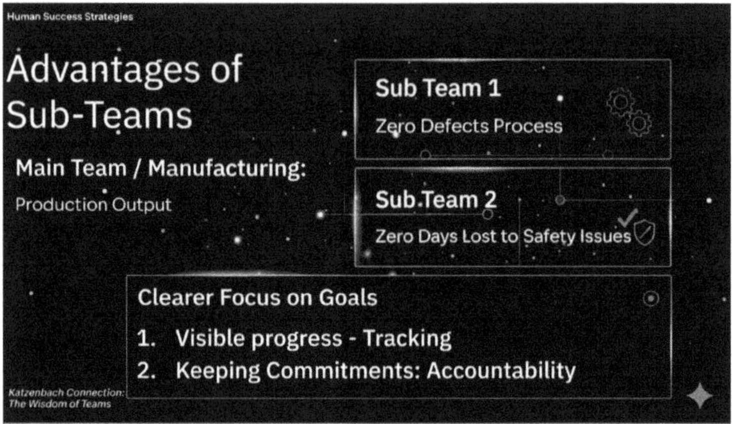

Evan began to see a new option here. He could harness the power of AI through the lens of sub-teams.

He would ask his reports to experiment, evaluate and embed new and specific AI tools into existing workflows and processes. The new sub-teams they would create would be tasked with identifying what apps and programs were already in use. The sub-teams would put emphasis on where, when and how AI was incorporated into the process.

And then, these teams would be asked to shift their focus away from the tools, with a new strategic question.

Teams, and sub-teams, would be committed to a common performance goal that is *greater* than the sum of individual or machine parts, he explained to Nina. Mutual accountability is enforced: the human team is accountable for the *final output*, including the AI's contributions, whatever they may be. Those contributions from AI would be subject to peer review and expert oversight.

He bounced these ideas off of Nina, sharing his thought process and going further into his explanation, "The 'team' now includes human members with complementary skills *plus* a well-defined relationship with the AI tool, which acts as a powerful but unthinking teammate whose work must be overseen."

Nina said, "I really like how you are seeing this," as she shared the next slide, which showed a thicker triangle. The sides were labeled, "Learning, Performance, Engagement". "What's this?" Evan asked.

"It's the framework for human success," Nina explained. "We have to help leaders develop their people, by giving them tools to understand what success means to their employees, and how these individuals can develop. Then, how to know, in real time, how people are doing. The idea here is more than just capturing sentiments, or check-ins, or wishes and feelings. We are assigning metrics for human success."

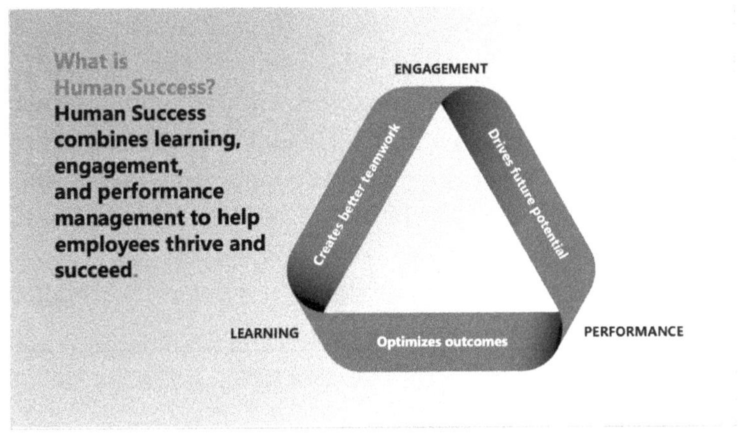

"What exactly does that get you?" Evan wanted to know.

"Well, for me, it's a seat at the table," Nina explained.

"Look, if revenues start to crash, the sales leader will call an 'all hands on deck' meeting, right? Because the numbers aren't there. Numbers drive the conversation. If finance runs low on cash, there's a similar reaction - because they can show numbers to explain what's missing..

"But we track numbers in our HR department," Evan said. "We keep an eye on attrition and training completion rates. We measure customer satisfaction and employee engagement."

"Right," Nina said. "But those are lagging measures. They are not predictive. Nobody really cares about attrition - it's just a characteristic of the business, like office furniture or something. I hate to say this but it's true. Some companies treat hiring and firing as an afterthought, in much the same way as you might purchase a new truck or get rid of an outdated laptop. That's a dangerous perspective when you're dealing with people," she said.

"A lot of executives and hiring managers say, 'We're a people-first company.' But that kind of talk is cheap. Employees and HR professionals just roll their eyes. Because they want to know if you can walk the talk - if you can prove it. If you can show that you really are focused on people. If you can, then you have numbers to back up that claim, or that cliché, or whatever you want to call it. Then, you set yourself apart. Now you can have a different conversation around attracting the best employees. Because you have hard data, not just clever sayings that everyone else is using, too."

Taking a pause to reflect, she continued. "The real data in most organizations - the HR numbers that matter - are either invisible, or at best: siloed and scattered."

Evan tilted his head to the side, considering the statement. He offered a slow nod in agreement.

What this little triangle means to me," Nina explained, "is having the same kinds of tools - the same kinds of numbers - for our organization. The difference here is that I have numbers for individuals - tracking progress around goals, training, and alignment, for example. But also, for the organization as a whole. It's a set of metrics that allows me to speak of the human element in the same way that finance might talk about ratios or inventory turns," she said, "or how IT might address service outages or service tickets in the current quarter." .

"So, you have a new set of numbers for human resources?" Evan said.

"Not exactly, but you're headed in the right direction," Nina said, turning from her laptop to look at Evan. "It's a new number for human *success*.

"There's one clear score to track and prove the impact of people on business performance, in real time. Now, this Human Success Score connects human resources data with business outcomes. It links engagement, learning and performance to growth and revenue. And it makes the connection at a micro-level, and a macro-level. So, leaders at every level are empowered to metrics that drive strategic decisions."

Evan had never seen this level of detail for monitoring individual progress and organizational health - not from a people perspective. What he was measuring today seemed somehow… incomplete. He was reflecting on his own conversations with his HR team - and when the conversations turned to sentiments instead of data, as they often did, it was hard to know what the real issues were.

In contrast, Nina's metrics were like a sleep score from a digital wristwatch: tracking areas of success as well as places for improvement, in real time.

On the screen, Evan saw a new slide, featuring the four key aspects of "Engagement":

- **Checking In** - Am I regularly checking in with my manager (ideally weekly) sharing my challenges, achievements and where I need support?
- **Reflection** - have I reflected on my successes, contribution to the business, and my challenges recently?
- **Recognition** - am I providing and receiving recognition from my peers on a regular basis?
- **Feedback** - have I received and given targeted feedback to my peers?

Evan wondered how these aspects of engagement might relate to the concepts from *The Wisdom of Teams*. Nina clicked back to another presentation, and Evan saw the answer:

Human Success Strategies.

By using dedicated sub-teams to establish governance, experimentation and job redesign around AI, organizations can channel the power of mutual accountability to ensure the technology augments human potential rather than replacing it outright. This makes the human team the necessary, irreplaceable "human in the loop" for high-quality collective performance.

ZENSAI

"The key here," Evan said, making a realization, "is where we focus. We have to keep our attention on human success," he said. He pointed at the screen, tapping at the words *"human in the loop"*.

The idea landed immediately for Evan, like the pigeon near his left foot. The plump grey bird was pecking at breadcrumbs on the cement.

"AI is not the hero of this story. And it's not the villain," he said.

"People are managing via headlines right now. The investments in AI are substantial, the capabilities are amazing, the news stories are everywhere. And with billions of dollars being invested or exchanged - you be the judge of that - it's hard to know if we are in a boomtown or a bubble. Doesn't matter, really. What I have to do, as CEO," Evan said, "is look beyond the hype. I need a strategy that's timeless, but still modern and relevant and fresh," he said.

Evan sighed. "Nina, I gotta tell you: one of the biggest parts of this job, and it is not always easy, is being able to separate signals from noise."

Evan recognized that whatever AI was doing, or whatever it would become, the market intention would always be around human success. The timeless wisdom here, he reasoned, was bigger than AI.

He told Nina that he couldn't build a sustainable business based on chasing trends.

"You have to be a person of your time," he said, speaking in general terms about anyone who didn't want to get run over by progress and technology. "You have to see the trends and recognize resources but always remember that there are pieces and elements of business that remain, as constants, even in a sea of variables and change.

"The conversation, for example. We are always having conversations, and engaging in dialogue, in order to find that alchemy that Robin was talking about yesterday. Sometimes those conversations are going to involve AI. And so, we need to give some awareness and some attention to how the conversation is changing. But also, we need to focus on the numbers in the narrative." He pointed at the screen. "That's what you're talking about, right here. Strategic guidance with measurable results.

"After all," Evan continued, "who knows where AI will be in a year or two from now? No one can guess. But here's my bet: heaven willing, humans like you and me will still be here. Companies will still need teams. And customers. And revenues. That's why human success matters.

"Maybe we are going to need to cut our workforce," Evan

said, standing. "We can always find ways to be more efficient and more effective. But sometimes that means that we need to hire, and build - not shrink. After all, growth requires investment - not contraction." he chuckled. He paused as his words regained a serious tone. "The only way to move forward with any business decision," he said, "is from a place of understanding. Not blind compliance."

Nina nodded. She appreciated his perspective, in much the same way that the CEO appreciated her own.

Evan thanked her for sharing the slides, and the insights into *The Wisdom of Teams*.

"I've been looking in the wrong direction," he confessed. "I've gotten caught up in the hype. Listening to the noise. With this board pressure stuff, I've been chasing a workforce reduction number. It's a misdirection," Evan said. "Yesterday Robin said, 'Inputs plus Outputs equals Outcomes'. I still have the ability to influence all three of those things."

He took a moment to consider that he didn't exactly know how, just yet. But there was something he was certain about: "Trend-hopping just isn't my style," he said to Nina. Then, in a voice just above a whisper, he said to himself, "But trend*setting* just might be."

He was already thinking about how to shift towards the teams and individuals inside his organization. He knew that they were the real and timeless heroes of the story. The strategy here wasn't about AI, as Lachlan had suggested. Evan's strategy required a deeper and clearer focus on the company's most important resource: its people.

———

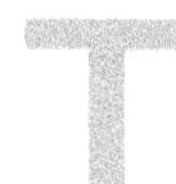

The Personalized Playbook

1. **Chasing the Misdirection:** Evan realizes he was "chasing a workforce reduction number" - which was a distraction. What single metric, trend, or competitor initiative is currently misdirecting your leadership team's focus away from building sustainable human and organizational capabilities?

2. **Timeless Wisdom:** *The Wisdom of Teams* is decades old, yet its principles remain vital. What is one fundamental "timeless truth" about people, culture, or organizational design that your company has recently neglected in favor of a newer, trendier strategy (like a new AI tool, or a fleeting organizational structure)?

3. **From Common to Uncommon Sense:** Katzenbach distinguishes between common sense and uncommon sense. Where is your organization making common-sense decisions (e.g., cutting training budgets during a downturn) that ultimately undermine the uncommon sense required for high-performance (e.g., investing in new skills for AI augmentation)?

4. **Trend-Setter or Trend-Hopper?** Evan commits to becoming a "trend-setter" in business governance. How can your organization use the adoption of AI not just to survive, but to create a new, superior model for human / AI alignment - something that your competitors will be forced to copy, perhaps? What does that "trendsetting" model look like for you, inside a single sentence?

"The org chart is the problem," Rasmus said.

He and Robin had returned to the plaza in front of the food truck, with Nina. Evan had seen them in the lobby, on his way back to his room after the deep dive with Nina. He said he needed to make some quick phone calls for the day ahead. He promised to come back to their conversation in just a few minutes. Nina, Robin and Rasmus sat on the benches that surrounded the plaza, for a quick debrief.

Robin was curious about Nina's conversation with Evan. She explained what they had discussed, and how the idea of sub-teams had captured Evan's imagination.

"Traditional hierarchies aren't nimble enough for the age of AI. Old org charts measure control - Human Success measures collaboration speed. Inside any company, there's one sub-team that isn't really a sub-team at all," Rasmus continued. "It's the main team - the main structure around Human Success."

"You're talking about The Trifecta," Nina said.

"But did you explain about the Human Success Score?" Robin asked. "With Evan, I mean?"

"We talked about the score in terms of impact and compared the numbers to what finance or IT or sales would use for a leadership discussion. But I think I need to go further, maybe. He seemed to get it - and he said that the level of measurement was different from what he was doing now," Nina said.

"But what about the consequences?" Rasmus asked. "What about the number?"

Nina and Robin knew what he meant. The number was the key to managing risk - and leading by example - inside of Human Success.

The danger, for any company, was in these two words: do nothing. The status quo was not going to work, especially when it came to leading an organization in the age of AI. Evan knew he had to do something different. Rasmus, Nina and Robin had already done it. But would it work for Evan - and for his Board of Directors?

The key was in the number, Nina knew. Rasmus was talking about how HR leaders - human success officers - needed the same kind of hire/fire consequences that were there for the Chief Sales Officer, or the CFO. Consequences that meant you either kept your job or lost it. Hire/fire was a powerful metric.

If the numbers weren't there, if performance wasn't happening, these roles were in jeopardy. If the company was $100 million behind in EBITDA and missing projections, the board wouldn't have to say to the CEO, "Hey, have you talked to your CFO?" Rasmus knew there would be no need for that conversation. The conversation would be, "The CFO is out. Period." And maybe even the CEO as well.

What did those consequences look like for the CHSO? What were the metrics of meaning, around the company's most vital resource (its people)? Success was invisible without some kind of metrics tied to CHSO performance. Maybe invisible was too strong a word? But Nina knew that what gets measured gets done. The conversation around attrition and other sentiments, from an HR perspective, often felt "squishy" and shapeless. That's why Nina had started her software company - to correct what was broken inside of HR. And to provide insights into the measurements that mattered.

The key performance indicators for the CHSO were expressed in the Human Success Score. The metrics were a way to know if the organization was really thriving - and if the CHSO was doing her job. That measurement was the number that mattered, for the entire workforce.

That's why the CEO, CFO and CHSO needed to form a targeted team - The Trifecta. A Trifecta that was focused on numbers and metrics. A Trifecta that had visibility into overall leadership, overall finance, and overall organizational health. These three positions, united in a single quest to refine and reform the organizational structure, was a requirement (not an option). This sub-team, made up of three top executives, formed the point of the spear, in organizational transformation: The Trifecta.

The next-level CHSO, Nina knew, had to be business-oriented. Numbers-focused. And accountable for the performance and development of the organization. But without the connection and visibility from the CFO, investment in the human aspect of the business would be murky. Unclear. Unfocused. And, for the CEO, engaging in The Trifecta allowed for greater insight and control over the human aspect of the business - from both a financial and operational perspective.

"The Trifecta has to focus on the Rule of 40," Rasmus said. "It may not be a perfect fit for Evan's business model, but it's one way to measure success - and it will resonate with the board. I'm sure of it."

He was talking about the measurement between growth and profit margins - the exact number that Lachlan was trying to influence with his 40% cut.

In general terms, the Rule of 40 is applied to SaaS (software as a service) companies in the startup stage. The term was popularized by Colorado venture capitalist, Brad Feld, as a method of measuring a company's operating performance.

The idea is that the sum of the growth rate and profit margin should be equal to or exceed 40%. The benchmark combines margin and growth rate into a single number, to help investors protect downside risk - and steer the company towards success over time. What Rasmus knew, or at least suspected, was that Evan's board was looking for EBITDA numbers that matched growth. But there were multiple combinations - multiple possibilities - that could all add up to the Rule of 40.

Brad Feld explained it like this:

The 40% rule is that your growth rate + your profit should add up to 40%. So, if you are growing at 20%, you should be generating a profit of 20%. If you are growing at 40%, you should be generating a 0% profit. If you are growing at 50%, you can lose 10% [in margin]. If you are doing better than the 40% rule, that's awesome.

"The key," Robin said, "is to get The Trifecta to focus on the number. Right? But Evan isn't running a SaaS company. What happens now?"

Rasmus replied, "It doesn't matter. The number for Evan might be 35%. Or 22%. Whatever. I don't need to know what it is. Because Evan does. Or he will," Rasmus shared. "The shift from pure revenue growth to bottom line focus is a combination of revenue growth plus EBITDA."

"And then," Nina said, "the people aspect will align to the business goals. Not some arbitrary number from the board, where people are laid off in large numbers, which might just destroy the company."

"That's right," Robin said. "Think of the signaling to Evan's marketplace, if he cuts 40% of the workforce. Even if that number is a slow roll over the course of a year, or two years, it's still going to look like the company is bleeding to death. What are layoffs going to do to their customer base, and their plans to grow revenue? You can't just say, 'AI will handle it', and expect the market to agree with you."

Nina and Rasmus nodded in unison. Robin continued, "Customers will look for alternatives from competitors that are healthier and thriving. Other players who are not removing their workforce in large numbers. Cuts like that are interpreted as signs of disaster, not growth."

"Growth is the key," Rasmus said, "in the context of EBITDA. That's what Evan's board really wants." He went on to emphasize The Trifecta as the key sub-team, referencing what Nina had discussed with Evan. Together with his CFO, Rasmus and Nina had already implemented The Trifecta. As the former co-founder of a software company, Nina saw the importance of tying finance to HR. So, when they discussed The Trifecta for the first time, Nina was totally on board.

But their CFO was not.

Jan (pronounced "Yan"), the company's CFO, dismissed The Trifecta idea as an attempt to bring relevance to HR - a relevance that he did not believe existed. He didn't want to cooperate - he saw the effort as a waste of time.

Jan, a straightforward Dane with a laser focus on the bottom line, had said, "How much does a 10-point increase in the 'Human Success Score' translate into hard dollar revenue, lower costs, or improved margin? This looks like a cost center initiative, not a profit driver."

But he didn't stop there. He had other concerns as well:

- He asked if "check-ins" and "training" metrics are truly leading indicators of organizational success, or just busywork. He was worried that subjective data will pollute the reliable, objective financial reports they rely on for governance.
- Like a good CFO, he immediately focused on the expense of the new system, the analyst time required to build and maintain the "Human Success Score" (the new KPIs), and the potential disruption to established financial reporting cycles.
- He pointed to "analysis paralysis", saying that the Human Success Score was just another dashboard of complex metrics that would dilute management's

focus - leading to more reflection, and less action, towards measurable goals.

"Plus," Jan had said in the initial meeting, "how are employees going to respond to being 'graded' and monitored in real-time? Are people going to game the system, and just input information around check-ins that didn't happen, so that they can improve their scores? How do we create trust inside this process, for the team?"

Rasmus had looked over at Nina at this point in the meeting. Jan had been running around, unbridled, and it was time for a counterpoint. Rasmus raised his eyebrows. With a nod, his expression said to Nina, "You're on."

Nina started off with the cost of attrition, and the ROI of retention. Jan began to realize that having an integrated "Human Success Score" provided a predictive model for turnover. With it, they could quantify, in real dollars, the cost of replacing an employee (recruiting fees, lost productivity, training costs) and directly link preventative investments (training, check-ins) to a reduction in that cost. This turned the HR budget from a cost into an investment with a measurable return.

"So, it's not just some random expense, or a pollution of data. The Human Success Score is the number we need," Nina said. She looked at Jan, as he made a note in his notebook. She wondered if anything was making sense for him.

She went on to explain how the number would impact workforce planning and capital allocation: the integration of HR and finance metrics allows the CFO to model future workforce needs with far greater precision. Instead of simply freezing headcount, or cutting it, they can strategically allocate capital to high-performing, high-success-score teams. This supports strategic resource allocation rather than reactive cost-cutting.

Now the company could maneuver with precision: cutting or adding based on objective criteria.

"And here's how it fits for AI," Rasmus said, backing Nina's play. "The AI age requires a clear understanding of which human skills are being augmented, which are becoming obsolete, and where new skills need to be developed. The Human Success Score's metrics on training and learning become vital. Everybody's talking about upskilling and every employee is wondering how to make sure they stay relevant. Training is the key to performance."

Nina remembered how Jan started to come around. It began with a look in his eyes, a moment where he saw what he always saw - the power of numbers - and applied it to what had traditionally been called, "human resources". He saw this new tracking data as the only way to ensure the company is investing its training dollars into the human skills (like critical thinking, creativity, and emotional intelligence) that are *most complementary* to AI, thereby maximizing the return on AI technology investments.

"There's a new team member," Nina said. "It's AI. We have to learn how to use the tool and coach others to make intelligent decisions around their contribution and around their careers."

But Rome wasn't built in a day. Jan's initial reluctance remained, but somehow he found a way to open up to possibilities. At first, he aligned around the numbers. Then, he turned into a supporter as the results started driving new behavior. Weekly meetings for The Trifecta really helped.

The team implemented the Human Success Score, and began adopting a new approach. After the first organizational analysis, Rasmus, Nina and Jan discovered that their Human Success Score was 66.

That's not 66 out of 70. That's 66 out of 100.

There was work to be done, to be sure. And, like everything in business, implementing a new and sustainable leadership initiative like this takes time.

Here's why the investment in Human Success was worth it:

- **Intangible Asset Valuation:** In the modern economy, human capital is the primary source of competitive advantage. Jan realized that for investment purposes and M&A due diligence, they need to place a reliable value on the company's intangible assets.
- **Balancing the Balance Sheet:** the company tracked IT investment, since the CIO had a dotted-line connection to Jan and the finance team. There were clear numbers around what the company was spending on AI. Why not have the same level of measurement and understanding around the performance and investment in people?
- **Creating Structure and Scale:** The Human Success Score provides a structured, repeatable, and defendable set of metrics that investors and board members can use to evaluate the quality and future-readiness of the workforce—the ultimate governance and structure tool for modern capital.

Over time, Jan moved from seeing The Trifecta as a veiled attempt at HR relevance. Today, Jan doesn't just talk about the costs of a project - he wonders about the burnout risk score as well. Because he sees a more comprehensive method of accounting.

Jan came to understand the Human Success Score as a vital number, seeing it as the only effective way to quantify, govern, and strategically manage the most important and most expensive asset in the company: its people.

In turn, Rasmus had a cohesive and concrete story to share with the board around his workforce capabilities, performance and prospects. He had, for the first time, numbers to back up what he was seeing - even though the initial numbers were not good! The numbers, Rasmus knew, might not be good - but they were definitely informative, and showed the team what they needed to work on.

Traditional metrics like engagement scores were now presented with new depth and color, allowing for the kind of positioning that's on every CEO's mind: namely, putting the right people in the right roles. Hiring to fill the gaps. And building high-performance teams.

Nina, Rasmus and Robin were working with The Trifecta - a powerful sub-team that was focused on greater output from their organization. Now they had to share their story with Evan - and see how it might help him. But first, the team needed to refuel on coffee. They stood up and took a place in line, as Evan walked out of the front doors of the hotel and approached them.

———

The Personalized Playbook

1. **Challenging the Org Chart:** Rasmus declares, "The org chart is the problem." What is the biggest organizational design flaw (e.g., silos, slow decision-making, redundant approval layers) in your company that is directly inhibiting the rapid adoption and effective utilization of AI?

2. **The Sub-Team Governance:** The Trifecta is a governance structure that acts like a sub-team. Identify one major strategic challenge in your organization (e.g., cross-functional product delivery, AI ethics, long-term talent retention) that requires a governance structure that exists *outside* of the traditional, hierarchical org chart. Who are the three executive functions that would make up this "Trifecta"?

3. **Data with Consequences:** Rasmus raises the question of consequences for the Human Success Score. If your managers' performance was tied to transparent metrics like burnout risk and collaboration speed, how would your management culture change overnight? Are you willing to embrace the discomfort and accountability that comes with this level of transparency?

4. **The Threat of the Status Quo:** The greatest danger is "do nothing." Identify one critical initiative that your organization has tabled or delayed due to internal debate or uncertainty about AI. What is the minimal, actionable step you can take this week to break the status quo and move that initiative forward, even without a perfect plan?

5. **From Skeptic to Champion:** The experience with Jan shows that financial skeptics can be won over by data. Which key internal stakeholder (e.g., CFO, COO, a veteran leader) is currently viewing the focus on "Human Success" as "veiled HR relevance"? What single, hard number from your organization would you need to show them to transform them from a skeptic into a champion?

CHAPTER 12

THE RISE OF THE HUMAN SUCCESS OFFICER

After the quick stop at the coffee truck, Evan jumped into a dialogue with Nina and Robin - talking about The Trifecta. He was intrigued by the Rule of 40, and was already considering how he might modify Brad Feld's formula to fit his own circumstances.

Meanwhile, Rasmus was writing something in a notebook. A specialist at multi-tasking, he was listening to the conversation as he sketched out some ideas for Evan. At a pause in their chat, he held up an image for Evan to consider.

It was a combination of triangles and diamonds, labeled in block letters. A large triangle was on the top of the image, and it was labeled "Old". Underneath it, a triangle with a diamond shape drawn onto at its peak was labeled "New". Looking at the image, Evan said, "What's this?"

"This is why companies need a Chief Human Success Officer," Rasmus said.

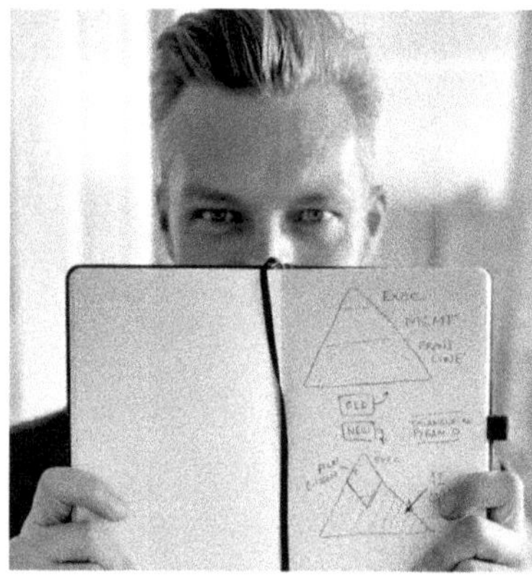

Rasmus shares his notes with Evan and Robin.

Robin recognized the drawing immediately. He'd talked about it on multiple occasions with Rasmus and Nina, so he picked up the conversation.

"That top triangle," Robin said, "is how we think of the traditional organization. The Exec level is the C-suite, made up of a small leadership team. Then, the management layer - which is already shrinking at a lot of organizations - is in the middle section of the triangle. And the 'Front Line' label stands for individual contributors and front line workers - which is where things get made, customers get service, and things like that."

Evan said, "What about this diamond shape, on the bottom triangle? What's that?"

"I apologize, it's probably a little unclear," Rasmus said. "I guess I'm more of a 'patio Picasso' than a real Rembrandt. But what this diamond is showing is the new organizational design. We still see the executive layer, and a layer of oversight, which you might call directors or supervisors or whatever fits for a particular organization.

"But instead of seeing a massive amount of workers on the front line, we are seeing a change. A shrinkage, if you will, so that the big triangle," he said, pointing at the top image with a black pen, "changes into a smaller diamond. There are not as many people needed for front-line work, such as customer service for example. Expansion comes from adding AI capabilities."

Nina added, "That new shape to the organization represents doing more with less people. AI-native companies utilize this kind of diamond-shaped org structure. Where there's a small team that's leading things, a layer of oversight for LLM tools and integration, and things like process flows and other aspects of production are handled by AI."

"Right," Evan said, pointing at what looked more like a kite than a diamond in the drawing. "This diamond is the shape of an AI-centric organization, where the revenues per employee are probably ten times what our numbers look like. So, what's the answer here for my situation? Cut 40% of the workforce and eliminate folks in customer service, or sales, or what exactly?" Evan wondered. Was this kind of AI expansion - where technology filled out the pyramid - what Lachlan had in mind? Evan wanted a better answer. He knew it was out there.

"Look, there are a lot of ways to make money," Rasmus said to his old friend, as he shifted back to the practical application of the concepts. "And many ways to design an organization. You know you can't just compress a thick triangle into a skinny diamond, bolt on some technology, and call it a digital transformation. Especially when that's not your path. At least, not yet.

"But maximizing outputs from your people, especially in a

world where native AI companies are attracting investments and customers, means *concentrating* on the people inside the organization," Rasmus added.

"That's why companies need to shift to this Trifecta we are talking about," Robin said. "Because it provides you with the competitive focus and financial insight you need. And, quite frankly, if you are asked to do more with less - less people, for example - you've got to make the shift from 'human resources' to human success.

"It's not just words or labels," Robin continued. "Human resources is an old way of looking at something that is actually becoming more valuable and more important in the age of AI. Because, if you're running an organization with less people, well, now it is even more critical to find the *right* people. And then, focus the organization on the thing that your customers and shareholders need the most: human success."

Nina said, "HR used to focus on people and culture, with perks like free snacks and Kombucha machines." Evan chuckled at the Silicon Valley reference. "We saw the office as a playground, and said things like, 'we're family', right? I mean, admin duties, like payroll, policies, and vacation tracking are important. But not necessarily *strategically* important. For me, I wanted to move beyond compliance. And I know that many other HR leaders feel the same way. There are a lot of human-first companies that are looking for ways to develop their people, and create a new level of success."

"And that success," Rasmus added, bringing the conversation back to EBITDA and critical numbers, "has bottom-line impact."

Evan looked down at his own notes. He still went old

school, like Rasmus, and captured things with pen and paper. The pen felt good in his hand - a connection that somehow deepened their conversation for him. He appreciated the link between the paper, his memory, and new insights, more so than tapping on a keyboard or thumbing notes into his phone. He saw his scribbles around the rule of 40, and how he planned to connect his CFO and his HR leadership. He had written, "Get this done!" on the page.

He had written those words. Not Lachlan.

"It's a shift," Evan said, "this Human Success Officer role. It's a framework. A framework that's moving from compliance to development, really."

Robin added, "And from reactive to proactive. So that the whole is greater than the sum of the parts it's made of."

Evan liked the sound of that.

His phone buzzed in his jacket pocket: a reminder of his upcoming team meeting. A new plan was already taking shape in Evan's mind. A fresh perspective that would draw a new battle line with Lachlan - because the numbers would tell the story. The rise of the Human Success Officer role would be the foundation of Evan's new initiatives.

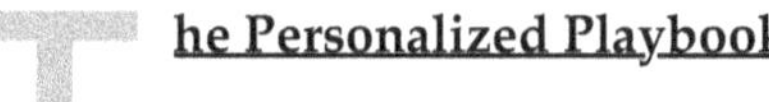

The Personalized Playbook

1. **Defining the Apex:** The old organizational chart puts the C-suite at the apex, funneling control upward. What is the single biggest problem caused in your organization by the current hierarchical structure, and how would a new structure focused on pushing capability and autonomy to the front lines help solve it?

2. **The CHSO's Mandate:** The Chief Human Success Officer is a shift from compliance to development, and from reactive to proactive. What are the three non-negotiable financial metrics (beyond attrition and engagement) that a CHSO in your organization would need to be responsible for, to prove their strategic value to the CFO and Board?

3. **The Rule of 40 Modification:** Evan is considering how to modify the Rule of 40 (Growth Rate % + EBITDA $\geq$ 40) to fit his circumstances. If you were to create a "Rule of Human Success" for your company, what would your formula be? (Example: Revenue Per Augmented Employee % + Human Success Score % $\geq$X).

4. **Beyond HR Compliance:** Evan notes the shift from compliance to development. Which of your current HR functions are primarily focused on compliance (checking boxes, mitigating risk), and which are focused on development (building future capabilities, increasing impact)? What steps can you take to shift the focus and budget toward development?

PART THREE
MAKING IT HAPPEN

CHAPTER 13
WORKING THE PLAN

A light rain was falling on the streets of London. Robin exited Paddington Station and began walking across the plaza to Microsoft's headquarters, a massive glass and steel structure near Westminster. Looking at all the miniature bear statues and artifacts in the station, Robin chuckled at the memory of watching the *Paddington* movie with his son, all those years ago.

Dodging the raindrops, Robin clutched the collar of his coat to block out the London cold. He entered the building through a large revolving door and looked around. The enormous lobby was bustling with people. An 11-story glass atrium marked the edge of the building; its angled windows spotted with raindrops. Across a sea of people that seemed to be moving in every direction at the same time, he spotted Evan Sato seated near an embankment that held plants and ferns. Robin waved at him as he walked towards the CEO.

The men exchanged greetings and sat down. It had been a month since the meeting in New York. Robin had agreed to meetup with Evan to talk about his progress.

Speaking via telephone last week, to make the arrangements, Evan had told Robin, "I've formed The Trifecta. It's good. We are making progress, definitely in the C-Suite. We're asking sub-teams to meet at least weekly, to examine how we re-architect workflows. But we've run into a problem."

"Like what? What kind of problem?" Robin asked.

"Do you know the story of the turkeys at Christmas?" Evan said.

Robin thought for a second. "That's the one where somebody asks the turkeys to vote on Christmas, isn't it? And when the turkeys realize what's coming for them, they decide to vote against Christmas. Is that the one - where Christmas gets cancelled?" Evan nodded in response.

"It's what happens when you ask employees to participate in a change that might lead to their own job replacement or removal," he shared.

Robin had seen the voting turkey dilemma play out on multiple fronts. He had read the story of an employee, Kevin Cantera, from Las Cruces, New Mexico, who worked as a researcher and historian at an edutech company. As part of new initiatives to embrace AI, Kevin went "all in" - becoming a sort of "AI Whisperer" at his firm, according to an article in the *Washington Post*. Cantera, a 17-year veteran of his firm, told the *Post* that "AI was an incredible tool for me as a writer. I considered the LLM as a collaborator," Cantera said. However, as *Futurism* magazine reported, "The LLM was a collaborator alright - a collaborator with his boss."

Cantera, along with a few dozen of his co-workers, was fired. And replaced by AI.

Inside their phone conversation, where Robin shared the story from the *Washington Post*, the two men realized they would both be in London soon. They agreed to meet up at Microsoft's Paddington headquarters. "There's a way to lead and govern with people-first AI," Robin said. "I'll explain more when we meet."

Inside Microsoft's enormous lobby, Robin asked Evan a question. "So is AI something that adds capacity or takes away resources?" he asked.

"Both," Evan said. He took a sip from a water bottle and wondered where Robin was headed.

"But in terms of a governance framework and a sustainable leadership model, the relationship between the workforce and AI must be expansive, right?" Robin said. "Instead of focusing on reducing headcount, which makes every 'turkey' nervous, consider emphasizing how AI will increase capacity. And reinforce the value of the work team, as well as the individuals who implement new solutions."

The look on Evan's face conveyed that Robin's words were easier said than done. Was this just spin, or was there substance here?

"Quantify the number of hours that are spent on low-value work that could be automated," Robin emphasized. "Combine that with a focus on new work: the result of freed capacity. For example, for marketing or product development teams, instead of summarizing reports, you can focus 80% of your time on audience insight and innovation. What's the value of that excess capacity? And if your teams aren't experiencing excess capacity with AI, they're doing it wrong."

"That's a bold statement," Evan said. "When you consider the study by MIT that found that companies that integrated AI saw no meaningful growth or revenue," he shared.

"Look, I've read the study - I know what you mean," Robin said. "In an AI-assisted cohort, less than 44% of the tips and suggestions from AI were actually implemented without modification," he said. "It was a very specific study, and I'm not sure that it's fair to draw the conclusion that 'AI doesn't work'. The key takeaway is that often our expectations are different than reality, especially when it comes to working with AI. From my point of view, the study is a call to action - there's never been a more important reason to focus on human success. What if the failure isn't in the AI, but in the way we design the workflows, the interaction, and the collaboration with the LLMs? That's a question of leadership and collaboration, as well as organizational design. It's too easy to say, 'oh, AI is broken'. The thing that's not working is the way that we design our organizations, and teams, so that we can create new outputs and realize new outcomes."

He has a point, Evan thought to himself. It seemed that the question that every CEO had to answer was: how are you going to work with AI? The tools were here. The way that humans were using them was where we needed to focus.

Robin continued, "You ask people to identify the tacit knowledge and relationship equity that goes into the completion of various tasks. You're asking teams to identify the creative problem-solving skills that AI *cannot* reproduce or replace. This aspect represents the human-only resource that must be paired with AI."

"I hear ya. But I'm still sensing a mistrust amongst the organization," Evan said, cautiously.

"I'm wondering about how you've set up the context," Robin said. "When I work with any CEO, or with my teams, I always work around a sense of urgency. A goal that needs to be met, of course, but also an obstacle - an enemy or a foe - that needs to be overcome. That story about the turkeys at Christmas isn't the best metaphor, because AI isn't the farmer's axe or the slaughterhouse. The key is to institute a formal policy that any customer-facing or decision-critical output is reviewed by a human subject-matter expert. That way, the human is the final quality and ethical resource in the system - signaling that human know-how, ethics and decision-making are at the heart of Human Success."

"OK, but what's one way to increase trust, given that some employees may very well be replaced in this process?" Evan asked.

"AI adoption is accelerated during a crisis, or challenge. Communicating the board's wishes is critical," Robin started. But Evan interrupted him.

"So, I just tell everyone that the board wants to cut the workforce by 40%? How does that work, my friend?"

Robin said, "I'm not sure that 40% is the number and neither are you. Look, we both see this the same way: the board has given you tactical advice on a strategic problem. What they want, as I understand it - and correct me if I'm wrong here - but what they want is a better bottom line, from a company that's doing more with less. Maybe that's less people. Probably it is. But you told us in New York that you were finding new numbers, and new equations, to give the board what they want. AI is part of the story. It's not the whole story. Ultimately, the business objectives are yours to own," Robin said. "And so are the ways you get there.

The next question surprised Evan. "What's the framing of the crisis or challenge that's not going to make people feel like turkeys at Christmas?" Robin asked.

Evan realized he hadn't really provided that context for the initiatives - a context of urgency and acceleration. AI, he knew, wasn't a driver for a "luxury cost-cut", but a necessary response to the existential threat of other nimbler, AI-enabled competitors. AI was a tool: a tool for competitive advantage, efficiency and expansion. But many viewed AI as a hatchet.

Evan had to find a way to maintain trust during this time of transition. AI was inevitable. But, as with any tool, the thing that made the tool valuable was in the way that you use it. Evan was sitting with that thought when he realized Robin was speaking.

"I'm here with Evan right now," Robin was saying into his phone. He held it in front of his face. He was having some sort of video call.

While Evan was lost in thought, Robin had FaceTimed Nina. Rasmus had told Evan that Robin's greatest skills weren't always what you see on stage, delivering a keynote. "He's a driver," Rasmus had told his friend. "Robin is better than almost anyone I know at getting things done, and keeping teams motivated and on task." The quick call to Nina was absolutely on-brand for Robin. Evan appreciated the quick action.

"We're talking about the challenge of trust, inside the AI transition to human success," Robin said, as he set the phone in front of Evan.

Evan gave a tiny wave as he saw Nina's smiling face. He explained how he was looking for ways to lead with trust, and integrity, towards greater AI adoption. One month into the plan, with The Trifecta well-established (they were meeting bi-weekly for at least 30 minutes to discuss new initiatives) Evan had questions about greater system-wide adoption and implementation.

After they spent some time filling in the details for Nina, she said, "One thing that you can do is to create tangible rewards systems and incentives." The backdrop of a white-walled conference room framed her face as she spoke. "Make sure that employees who successfully collaborate with AI and tangibly increase team capacity are rewarded in some meaningful way. That can be with euros or dollars, via compensation, but there are many ways to recognize team members and teams for their alignment and adoption of new initiatives."

She continued, "Look for impossible victories."

"What do you mean, 'impossible victories?'" Evan said.

"Highlight and emphasize the ways that integrating AI has resulted in things that seemed impossible before. Get people talking about what happens when hours are put back into the day, and how shifting from tasks to strategy changes the way work gets done. Emphasize what's happening that's moving business forward. Begin socializing stories of how people are creating the impossible and finding victories and wins inside every part of the organization."

Robin looked away from the phone and faced Evan. "That's the communication aspect of what you're doing. You are being transparent and showing everyone what progress looks like. These initiatives aren't leading people to the slaughterhouse. They are igniting new growth and sparking new profitability. That's the story you have to share, not just in words but in action. Keep mapping workflows, not tasks," Robin said.

Evan recalled how Robin had shared the importance of focusing on outputs to create outcomes - a process that made total sense to him, and something his company was already exploring on a deeper level.

The conversation continued via the FaceTime call, as Evan shared several ideas for celebrating wins and recognizing progress - with benefits such as additional time off or other perks. "Whatever you can do to encourage the sharing of prompting mastery, for example, is key," Nina said, reinforcing Evan's ideas, "The idea is to make sure that you incentivize people and teams to embrace what Katzenbach said, and then to apply it."

Indeed, Evan began to see what his teams needed to embrace: namely, that AI was enabling the organization to survive. With new tools in place, new profits and efficiencies were already emerging. Collaboration - humans and AI working together - would allow the company to conquer new markets.

AI was a survival tool. If business were a game of poker, AI would be table stakes. In other words, you had to have it in order to stay in the game. You had to have table stakes in order to make big bets in the game. And win.

Evan knew that AI was key to securing the overall future of the company, and the future of its employees. Would people need to be cut? Probably. Evan couldn't change that simple fact. No one could. And he wasn't on some quest to save everyone. Or shed a tear over what might be an inevitable outcome. Evan was a realist.

He was a CEO. His job was to maximize shareholder value, and meet the goals of the organization, by whatever means necessary within the boundaries of legal and corporate constructs. He would do what needed to be done, as he had done countless times before, when people needed to be let go.

But, as he had learned in his father's woodshop, he wanted to work carefully so that he did not cut too deep. He knew that pushing back on Lachlan's overreaching input was going to be part of the game. For Evan, how he approached a problem was vital to how he solved it. Just as with a word problem in Algebra class: the way you set it up determines your result. He wanted to have his inputs and his variables in place, so that he could show a solution to Lachlan (and the entire board) that reflected what they really needed from the company.

His ability to strategize and implement the plan would need to bring real results, real quick, if Lachlan were going to listen.

Turning the redesign of the organization into a team effort seemed transparent and logical - and a lot smarter than asking turkeys for a vote.

"Can you access your email?" Nina said, noticing the corner of Evan's laptop in the frame on their Face-Time call. "Yes," Evan replied. "It's right here."

"Let me send you something," Nina said.

A few moments later, a message arrived from Nina. Inside a Word document, entitled, "From Replacement to Expansion - the AI Capacity Conversation" Evan saw the following:

Leading and Governing with People-First AI

To ask employees to form sub-teams and redesign their workflows without feeling like they are "designing their removal," leaders must implement a governance strategy that shifts the organizational focus from Task Automation (Replacement) to System Augmentation (Redesign).

- ### Shift the Conversation from "Replacement" to "Capacity"

The most crucial step is a communication and resource-mapping effort that changes the unit of value from "the individual employee" to the "capacity created by the human-AI system."

- *Governance Strategy*
- *Action for Leaders*
- *Resources to Identify*
- *Value Proposition*

1. *Stop saying: "We are implementing AI to reduce headcount/costs."*
2. *Start saying: "We are implementing AI to give our team **2,000 extra hours of capacity** per week, which will be re-invested in **new, competitive initiatives.**"*

*The "Hidden Capacity" Resource: Quantify the number of hours currently spent on repetitive, low-value work that can be automated (the areas where AI can contribute and make a difference). The time saved is the **resource** that funds the new strategy.*

- *New Roles & Focus: Clearly define and name the **new work** that the freed capacity will be dedicated to. This work must be strategic, joyful, and impossible to do before. Example: "Instead of summarizing reports, you will now focus 80% of your time on **Audience Insight & Innovation**.*
- *The "Expertise-in-the-Head" Resource: Identify and inventory the tacit knowledge, relationship equity, and creative problem-solving skills that AI cannot replicate. This is the **human-only resource** that must be paired with AI.*
- *Team Mandate: The mandate for sub-teams should be to design the **Human-AI Handoff Process,** not the code. Their task is to define the **new standards of quality** and where human judgment **must** be inserted to review, edit,*

and rewrite, ensuring no "workslop" reaches the customer.

- **The "Quality Assurance" Resource:** *Identify the subject-matter experts (SMEs) whose judgment is the **critical resource** for verifying AI output. These SMEs are the new "AI Editors" and their role is one of elevation, not elimination.*

2. A Governance Framework for Maintaining Trust

Leaders must establish a new governance framework that makes the relationship between AI and the workforce transparent and collaborative.

A. The "Human Veto" Rule

Trust Builder: *Institute a formal policy that for any customer-facing or decision-critical output, a human subject-matter expert retains the final right to review, modify, and veto the AI's output.*

Resource Management: *This explicitly identifies the human as the final quality and ethical resource in the system, ensuring that the critical know-how is still valued and applied.*

B. The "Productivity Share" Model

Trust Builder: *Ensure employees who successfully collaborate with AI and increase team capacity are tangibly rewarded. This could be through bonuses, promotions into new strategic roles, or dedicated time off. This breaks the Cantera paradox by ensuring high performance with AI is rewarded with advancement, not termination.*

Resource Management: *This encourages the sharing of prompting mastery and tacit knowledge, transforming it from a personal asset (hoarded knowledge) into a team resource (shared best practices).*

Evan looked up from his laptop screen. "This is perfect," he said to Robin. While Evan was reading the document, Robin had concluded the call with Nina. "Please tell Nina, 'thank you', from me," Evan said.

He continued, "I've got to excuse myself; I need to head over to Broadcast House. I'm doing an interview on global US business sentiments, with Tanya Beckett on *Talking Business.*"

"At the BBC? Impressive!" Robin said. Evan shmugged - that's what it's called when you smile and shrug at the same time.

"And then, when that's over," Evan shared, "I need to revise my plan!"

The Personalized Playbook

1. **Facing the Paradox:** Do your employees currently feel like the "Turkeys at Christmas"—rewarded for making themselves more efficient, only to fear their reward is eventual replacement? What is one specific question you can ask your team today to honestly gauge their level of fear regarding AI-driven efficiency?
2. **The AI Editor:** The human role is reframed as the "AI Editor"—elevating, not eliminating, the human. For your most critical, high-volume task currently being augmented by AI, how would you rewrite the job description to focus 80% of the human's time on editing, elevating, and ethically vetting the AI output, rather than initial creation or data input?

3. **The Human Veto Rule:** The "Human Veto Rule" builds trust by ensuring human oversight on critical outputs. Where is the most critical juncture in your customer-facing or decision-making process? Consider where instituting a formal, accountable human veto would immediately increase client confidence, build employee trust, and mitigate legal or ethical risk.

4. **Productivity Share Model:** The "Productivity Share Model" ensures that the financial gains from AI augmentation are tangibly shared with the employees who made it happen. Beyond a typical bonus structure, what is one creative way (e.g., dedicated strategic time, stock options, certification programs) you could use the financial efficiency gained by AI to advance the careers of your top AI-augmented employees?

5. **Trust as Governance:** The chapter argues that trust is a governance structure, not a soft skill. Identify one area of low trust in your organization (e.g., between sales and product, or management and front-line staff). What is one measurable, non-negotiable rule or policy you can introduce to structurally mandate transparency and accountability, thereby building trust?

CHAPTER 14
A COURSE CORRECTION

Rasmus opened the email from Evan.

"Rasmus,

We've been approaching human success all wrong.

We've made a category error. My CFO caught it. We are treating the transition to more AI like a standard IT project. Like an ERP or cloud migration.

Some of my peers have created AI task forces or hired Chief AI officers. But the results are mixed at best. Mixed? Maybe "missed" is the word? IDK. We aren't going to turn to IT and say, "Go figure this out." We aren't going to turn to HR or any department or individual in isolation. Isolating AI doesn't seem to be the answer. Engaging humans - all of them - around the usage and implementation is key.

AI is not just another tech tool. This is not like giving accoun-

tants calculators or bankers Excel or designers Photoshop. The work is human-centric, if we do it right. Execution is becoming abundant - in the past, execution was just expensive.

So, how we execute - which is ultimately a human decision and a strategic one - is really the issue here. Who are we, as an organization, when the capacity to execute is less expensive, and more expansive, than ever before?

Unfortunately, the teams are looking at this initiative as a one-off, much like moving to an ERP system or implementing one-time customizations in Salesforce. We have to understand how to commit to change and continuous disruption, as AI's capabilities continue to increase and improve. Making a one-time [massive] cut and calling it a win is not the answer here.

When I met with Robin and Nina in London, I realized something. It was something Robin had been trying to tell me since New York.

The new bottleneck isn't at the operational level, it's at the strategic level. We have to design entirely new human-AI workflows. In the past, our goal would have been to automate human tasks and achieve efficiency. Now, what we are realizing is that we need to re-architect work.

CIOs (and even CEOs) can't drive transformation by themselves. Only a business leader - someone with deep and intimate knowledge of workflows and key players - can separate the good stuff from the old ways of doing things. We are looking at legacy systems and long-standing processes and saying, "How can this thing be replaced?" We are not trying to throw everything out and replace it with some monolithic software, like an ERP implementation. We are trying to get people to identify and articulate business logic, so that we can explore how processes might transfer into an agentic system.

We are bumping into something called "productivity anchoring". I'm sure you've seen this: it's where people dig in on their value and cling to the hard way of doing things, because that's where they derive their own value and productivity. So, embracing AI for some is a conversation about identity and how they fit into the work process. We are working to re-think those anchors, so that the whole ship can move forward.

Rasmus stopped to reflect on the idea of productivity anchoring.

He had worked hard with his own team to provide the kind of mission-critical infrastructure that made the transition to the world of AI more seamless, more possible, and more real. AI can strip away friction in an organization's operations. The challenge is: focusing on human success so that productivity anchoring doesn't inhibit acceleration and growth.

For his CFO, Jan, and his entire C-Suite team, Rasmus knew that he had to find a way to turn anchors into enablers, and shift gatekeepers into facilitators, in order for progress to occur. The same expectation permeated every team in his company, by design.

Ultimately, the human success game was tactical, but also psychological. How people see themselves, and their contribution, has changed. And Rasmus knew it would only continue to change, as capabilities and advancements continued for LLMs. Culture, Rasmus believed, was really the overall state of mind of his company. And, in order to improve or shift that state of mind, he had to consider how teams were approaching the human element inside every task. Every part of the organization, including the people inside it, were there to drive numbers, growth and results. He turned back to the email from Evan:

Nina and Robin challenged me to look for impossible wins: where we are doing things now that were previously impossible with humans alone. We are doing things with massive-scale personalization, for example, in the way that we handle our marketing, outreach and sales. That's just the tip of the iceberg. We can't delegate transformation. We have to drive it.

Already we are seeing amazing progress. And we are also seeing the need for oversight and human intervention - exactly as you predicted.

I am attaching our 90-Day Leadership Plan. Would love to get your comments in Copenhagen next week,

ES

Rasmus clicked on the attachment and read Evan's plan. It opened with a comparison that set the stage for what he wanted to create:

**C**omparison of Human Success Initiatives around AI and Prior Digital Transformations

- **Leadership Differentiator**
- **Past Digital Transformation (e.g., Cloud, ERP)**
- **Leadership in the Age of AI (Human Success Focus)**
- **The Pacing**

Explore incremental change; build multi-year roadmaps to digitize existing processes.

- **Radical, Continuous Disruption**: The pace of change is the disruption itself—leaders must commit to constant, year-over-year re-invention.
- **The Governance**: Delegated to the CIO/IT as a technology implementation project.
- **CEO-Led Mandate**: AI is the most important strategic part of a business leader's job; it must be driven top-down.
- **The Bottleneck**: Execution. Build the software and manage system integration. Where can AI assist and reduce delays?
- **Strategic Design**: AI makes execution *abundant*. The new bottleneck is defining the right strategic business problems and designing entirely new human-AI workflows.
- **The Goal**: Automate existing human tasks; achieve efficiency.
- **Re-Architect Work**: achieve outcomes that were *previously impossible* with humans (e.g., massive-scale personalization).

PHASE 1 (Day 1–30): The Governance & Strategy Reset.

- **Goal:** Elevate AI from an IT concern to a core CEO and Board-level strategic function.

1. **Shut Down Delegated AI Projects**: Stop treating AI as a technology implementation. Immediately halt all fragmented, siloed AI pilot projects that lack strategic, CEO-level oversight. This stops the bleeding and prevents further organizational damage and wasted spend.

2. **Form the CEO-Led AI Action Committee**: Establish a cross-functional governance body (CEO, COO, Head of Product, Head of HR, General Counsel) to meet at least bi-weekly. This signals importance to the entire organization and ensures that decisions—on risk, ethics, and investment—are made efficiently at the top. **Comparison:** This replaces the past model of a *Steering Committee* which simply monitored progress; this new body *drives* transformation.

3. **Mandate AI Literacy at the Board Level**: Ensure the Board and executive team can ask fundamental questions without fear of judgment. This is crucial for overseeing risk and approving investments in a domain where lack of knowledge can lead to category errors and significant $ in losses.

4. **Define "The Unthinkable" Use Cases**: Shift the focus from automating low-value tasks to identifying 2–3 mission-critical problems that were impossible to solve with human labor alone (e.g., hyper-personalized customer journeys, real-time risk modeling). This harnesses the abundance of AI execution.

PHASE 2 (Day 31–60): Re-Architecting and Control.

- **Goal:** Design the new operational infrastructure and establish governance guardrails.

5. **Redefine the Role of the CIO**: The CIO/IT department must transition from **Gatekeeper** (limiting tools, enforcing rules) to **Enabler** (building secure infrastructure, creating enterprise-grade tool access, and maintaining AI observability).

6. **Implement "Accuracy First" Guardrails**: AI's greatest risk is the potential for hallucinations and

errors. Establish a clear policy for data security and accuracy.

7. **Map Workflows, Not Tasks**: For high-value, exception-heavy use cases (e.g., claims processing, complex customer support), the organization must manually map the undocumented "know-how" that resides in employee heads. Generative AI *challenges the concept of done*, so leaders must codify the operating procedures that AI will inherit.

PHASE 3 (Day 61–90): Culture Transformation (People)

- **Goal:** Secure employees' buy-in and treat AI adoption as a cultural, not technical, challenge.

8. **Address Displacement Head-On**: Acknowledge the natural fear (the "turkeys voting for Christmas" dilemma). Treat AI as a catalyst for upskilling and value-shifting, not just headcount reduction. Communicate that jobs are not being eliminated, but the *tasks* within them are—freeing employees for high-joy, high-value work (strategy, audience development, insights)

9. **Launch a Cross-Functional AI Literacy Program:** Mandate training for all employees on *how to work with* AI—not just how to use a tool. Focus on prompt engineering, critique, and verification, since the new work is about quality control and directing AI output.

10. **Institutionalize Continuous Improvement:** Unlike traditional software with clear release cycles, AI requires continuous monitoring and retraining. Establish a feedback loop where business users are empowered (with low/no-code tools) to contribute subject matter expertise, system integration updates,

and process improvements/ ongoing basis. AI is never "done".

Note: Security is paramount. No sensitive corporate data in public consumer models. Truth/Brand Control: Integrate AI models with internal, verified company data and knowledge graphs to dramatically reduce hallucination and ensure brand voice and compliance.

Rasmus sat back in his desk chair and smiled. The afternoon sunlight was streaming into his office, and he looked out at the canal. He knew his friend was well on his way to finding a new strategic plan and implementing it.

But would he be able to do so in the time he'd been given, by Lachlan and the board? How could Evan accelerate his implementation? Rasmus would find out, in the Copenhagen office, in just a few days.

The Personalized Playbook

1. **Defining the Category Error:** Are you currently treating AI as a fixed, one-time ERP-style project, or a continuous, never-ending investment in human capability? What specific line item in your current budget would you need to move from Capital Expense (CapEx) to Operating Expense (OpEx) to structurally commit to AI as continuous change?

2. **Commitment to the Infinite:** Evan realizes that AI is "never done". What is the one visible, high-impact initiative that your leadership team could institutionalize (e.g., a mandatory, monthly "AI Skill Share" session for all VPs) to signal to employees that continuous learning and adaptation is now a core function of every job, not an optional HR program?

3. **The Organizational Identity:** Evan asks: "Who are we, as an organization, when the capacity to execute is less expensive...?" What is your organization's answer to this question? Is your identity based on efficiency and cost-savings, or is it based on a unique human capability (e.g., trust, strategic insight, ethical judgment) that AI augments but can never replace?

4. **Countering Isolation:** Evan saw the danger in isolating AI. Which critical function (e.g., finance, product, sales) is currently implementing AI with the least amount of cross-functional oversight and collaboration? What is the immediate, low-cost structure (like a weekly "Trifecta" sub-team meeting) you can put in place to break that isolation and integrate learning?

5. **The Cross-Functional Mandate:** The 10-Point Plan includes a Cross-Functional AI Literacy Program. How would you structure this program to ensure that the required training is focused on prompt engineering and ethical critique (skills for the AI Editor) rather than just teaching employees how to click buttons in a new piece of software?

CHAPTER 15

THE REHEARSAL

The afternoon sunlight touched every corner, desk and shelf inside the Copenhagen office lobby. Tall white windows welcomed visitors into a sparse but comfortable waiting area, where Evan would soon meet with Rasmus. Evan sat in an oversized black armchair, part of the minimalist design of the lobby, and took in the surroundings. Curious about a plaque on the wall, Evan stood up and walked across the room to read the details: "Always Look for Uncommon Sense" was what it said. As he glanced around the lobby, he considered what he wanted to share with his friend.

As Rasmus had told him, inside his email reply, human success was an initiative that was designed to build real results. Based on recent information, and some conservative projections, Evan was already seeing what Rasmus had told him he needed: the number. The number that would prove the impact of human success. The number that would introduce a sustainable method of leadership and collaboration for the world of AI. The number that would be his ammunition in the fight for his company, his people, and his career.

The number, for Lachlan, was a headcount reduction. For

Evan, the number was something more. Something more nuanced. Something more multi-faceted. A number where the whole was greater than the sum of the parts it was made of. A number where AI was a welcome team member, but human success required more strategic thinking and more real-world insight from his team than ever before.

Evan wanted a different story around the number. For him, the number was an efficiency increase that punched up top line revenues as well. It was all interrelated, not isolated.

Human success, as they were putting it into action, was already creating new opportunities and transforming key metrics. The new structure, which Evan would soon share with Rasmus, had improved leadership communication throughout the company. But, Evan wondered, would it be enough? Would new numbers satisfy the board?

Profitability, Evan reasoned, was hard to argue with. Combined with growth and alignment around scalable results, Evan was building a business case for the new structure of human success. The structure was theory put into practice, with measurable results to reinforce the decisions that were happening not only in the C-Suite, or The Trifecta, but inside every team within the organization.

"You made it!" Rasmus said, appearing in the lobby. Evan turned around to face his friend. "Welcome!" The two men walked into the main offices, where Rasmus offered a quick tour and brief introductions to some key team members.

Returning to Rasmus' office, the men parked themselves at a conference table that sat against the back wall. On the other side of the room, on the shelves behind Rasmus' desk, photos of his family stood like soldiers at attention, guarding the proceedings in the CEO's base of operations. There were smiling faces of his family on ski slopes, in front of the Spanish Steps in Rome, and other similar locales. Mementos of Brentford FC, Rasmus' favorite football team in the UK – what Evan would call "soccer" - punctuated spaces between books and small glass trophies. A yellow ticket stub, about the size of a playing card, was inside a shadowbox frame. The score, "2-0", was written in large bright blue letters below – a reminder of an important victory in the Premier League for the team.

Closer to the conference table, beside the desk, a bigger frame held an award for Best Places to Work. The recognition dominated the side wall beside the conference table. Below it, a large photo of all the team members wearing orange jackets served as a reminder of who had made the award possible. Both frames and images were the same size, creating a sort of one-two visual punch for anyone who entered the office.

"Here's what I'm thinking," Evan said, looking at his laptop to get his bearings. Before he could continue, the door to the office opened.

"Hello, hi, I'm sorry I'm running a little late," Nina said, entering with arms full of paper and a laptop. She greeted Evan and settled in on his left, grabbing the third seat at the table.

Evan connected to the WIFI and projected a PowerPoint onto the television screen above the conference table. The large monitor hung on the wall to his immediate right, above a light tan credenza, offering an easy view of his screen for everyone.

"So, this is what you are wanting to share with the board?" Rasmus asked.

"Yes," Evan replied. "Let me just run through it like a rehearsal and stop me anytime something doesn't make sense." He cleared his throat and began the presentation.

"This restructuring plan is designed to leverage AI to create sustainable leadership while improving both the top and bottom line. We have already shifted our strategic focus, and will continue to do so, to drive measurable efficiency throughout the organization."

Rasmus interrupted him. "Where's the part where people get fired?"

Evan laughed. "Very funny. But also, not," he said.

"With your permission," he continued, adjusting his tone in much the same way that he would deal with an unwanted interruption from a board member, "I will outline our comprehensive approach so that you see the entire picture – not just a single number or headcount reduction."

Rasmus nodded in agreement. "Fair enough," he said, with a wry smile. Nina shook her head at her CEO's cheekiness and turned to look at the screen.

"We call this our Human Success Playbook," Evan said.

"The plan calls for a new governance model – a sustainable leadership plan that prioritizes human-AI collaboration, value creation, and overall process management, based on new metrics that we have introduced."

Rasmus had decided to lean in on the role of devil's advocate. "What's the number?" he asked.

Rasmus was acting like an impatient activist investor – good practice for when Evan had to face off with Lachlan. Evan sighed and chuckled under his breath at his friend's impertinence. He decided to stand up.

Just one week prior, on that rainy day in London, Robin had given Evan a piece of advice that he could not forget.

"Numbers," Robin had said to Evan inside Microsoft's headquarters, "are points in a narrative. The context is everything, when it comes to numbers. If your sales grow by 30%, that's a great story. But if your costs increase by 40% to get those results, you bury the lead. That's the most important part of a story. The context brings clarity. Numbers in isolation are incomplete. People may think that they want numbers, but what they need is the narrative."

Evan agreed, and he finished Robin's thought by saying, "Otherwise, you look like you are either hiding the ball, or you don't know all of the details."

Robin nodded in agreement. He said, "Think about it. Every result, every output, has a cost. And cost-cutting can have consequences, if handled poorly. Make sure you set the context for the numbers, otherwise the conversation or the report or the overview or whatever is incomplete."

"You have asked me here to share our plans," Evan said, returning his thoughts to the office in Copenhagen and the mock presentation for Rasmus and Nina. Once again, he assumed the tone he would take if he were in front of his board - which was to say his words became more focused. "The numbers you need exist within a context. I want to provide an understanding of that context so that we are all able to make informed decisions with complete information. I want to discuss our direction as well as our results, around human success, so that you can see that we're focused on both the bottom line and the future of the organization."

Rasmus said, "That's pretty good. Did Robin talk to you about context for numbers?"

Evan smiled and nodded once. Nina looked at Rasmus. The three leaders were on the same page. Robin wasn't in the room, but his ideas certainly were.

"Moving on," Evan said. "I want to start off with a new initiative from within one of our key sub-teams, which we call The Trifecta." An image flickered on the monitor, showing the connection between the CEO, CFO and CHSO, with the words, "Chief Human Success Officer" spelled out below that acronym. He elaborated on the new relationship and said that sub-teams were being formed throughout the organization.

"As a result of our renewed focus on human success, a key initiative has emerged which I want to share with you in detail."

He advanced the slide. The headline read: "Introducing the AI Control Tower."

Nina and Rasmus looked at each other, as Evan paused.

Rasmus sat up in his chair and propped his elbows on the table. He clasped his hands underneath his chin, mimicking the shape of The Trifecta triangle. He rested his head on his interlaced fingers, hiding his smile.

Nina leaned to her right and spoke in the general direction of Rasmus' left ear.

"I did not see that coming," she said, her voice just above a whisper.

Rasmus wondered if his reply to Evan's email had generated any new ideas. Evidently it had.

"May I continue?" Evan asked.

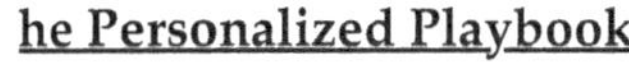

The Personalized Playbook

1. **Defining your story:** Lachlan's number was a story of reduction; Evan's number is a multi-faceted story of augmentation. What three key, non-financial data points (e.g., leadership capacity score, cross-functional speed, employee-led innovation rate) would you combine with a financial metric to create a nuanced number that tells a story of future value, not past cost?

2. **The Control Tower:** Evan introduced the AI Control Tower to centralize human strategy over decentralized AI execution. Where in your organization is AI currently being implemented in isolation (in a single department or as an IT project)? What specific cross-functional governance team must you create today to ensure that AI adoption is aligned, ethical, and continuously learning across the entire business?

3. **Rehearsal for Conviction:** Identify the single highest-stakes presentation or conversation you have coming up. What is the one non-negotiable **core belief** about human potential that you must communicate, regardless of the resistance you face?

4. **The Three Pillars:** The Trifecta (CEO, CFO, CHSO) is the core governance structure. If you were to create your own Trifecta, which three executive roles would you mandate to meet weekly to align human capital, financial investment, and technology

strategy? What authority would this Trifecta need to move beyond discussion to decisive action?

5. **Shifting to Offense:** What is the single, boldest, "trendsetting" initiative (as opposed to a common "trend-hopping" action) your company could launch this quarter that would force your competitors and skeptical investors to acknowledge the power of your human-centric model?

CHAPTER 16
LONG BEACH

"Value creation is our focus," Evan said, as he stepped away from the television monitor and smiled at his Board of Directors. "What you will see today is a nuanced strategy of value creation, sustainable efficiency, and margin expansion. Our new results are driven by our Human Success playbook - and strategic AI integration."

Inside the hotel conference room, the board members were seated in front of him. It had been almost eight months since the initial ambush from Lachlan, where Evan was given an edict to cut 40% of the workforce. The six months that Evan was given in New York turned out to be eight, in anticipation of the Annual Group Meeting (AGM).

The board filled eight of the 12 chairs at the conference table. Evan's CFO, Chief Sales Officer, and CHSO were also in the room.

The conference room for the Board Meeting.

The Hotel Maya in Long Beach was the site of the meeting, not too far from the company's headquarters in Douglas Park. Evan's team chose the hotel for its location: it's a short Uber ride for the board members who wanted to tour the offices before the meeting and see the new warehouse facility nearby. The proximity to Long Beach Airport was also a plus, as several board members had previously groaned about commercial travel through LAX.

The Hotel Maya, a 50-year-old property right on the water, had recently undergone a significant update to its rooms and facilities. The location, with clear views of the beach and the Queen Mary, provided a stunning backdrop for the meeting. The Latin-inspired atmosphere offered an executive retreat feel, helping to break down any potential board hostility. Or so Evan had hoped.

Inside the Luna Solstice conference room, the curtains were drawn shut for the CEO's PowerPoint presentation. In front of each board member, on the long table, was the Board Book: the core document containing the full presentation deck, financial statements, and supporting analysis for everyone in the meeting.

"The numbers tell our story," Evan said. He looked directly at Lachlan. "Because we are not managing via headline."

Lachlan didn't notice. He was looking down at his phone.

While the other board members were listening with rapt attention, Lachlan was scrolling.

Lachlan looked up. He offered a micro-shrug back to the CEO. His body language gave this headline: "Whatever."

Rasmus and Nina had told Evan that he should expect disinterest from Lachlan in the meeting, until the investor saw a number that lit his fire. That might be how the war begins, Rasmus said. A number would be his weapon of choice.

"You're looking for the 'hold my beer' moment," Nina said.

"What do you mean?" Evan asked, in their rehearsal in Copenhagen.

"It's something that some people say before they go into a bar fight, or they go hands-on and get something done. It sorta means, 'challenge accepted'. But also, 'hold my beer' is the first step before I can mix it up with this person or this situation or whatever," she explained.

Rasmus finished the thought: "'Hold my beer' is a prelude to a fight."

In Long Beach, Evan wasn't holding a beer. But the metaphor kept him on the lookout. And ready for a fight.

Evan advanced the slide and looked down at the long

table. He noticed that Lachlan had opened a thick leather-bound portfolio next to his Board Book.

That's odd, Evan thought to himself. He had been focused on kicking off the presentation. So he hadn't noticed Lachlan's conglomeration of dog-eared pages inside the notebook. Why did Lachlan pack an "extra" folio for the meeting?

While other board members were following along in the Board Book, Lachlan's remained closed. In an instant, Evan realized why the large portfolio was open. The papers inside contained Lachlan's rebuttal. He had come to the meeting with supporting documentation, and that was what was sitting in front of him.

Evan's next slide was another warning shot, aimed directly at the activist investor. Of course, the whole presentation wasn't designed around Lachlan. But this slide was. "One key metric here," the CEO said, "is *RPE* - revenue per employee."

The slide showed the acronym and prior versus current revenue per employee. "Instead of just looking at the number of people on the payroll, we are looking at the productivity and strategic alignment of those people, using AI."

"Aye," Lachlan said, his voice a low grumble that interrupted Evan's train of thought. "RPE. Got it. You're moving the goalposts here, Evan. Revenue is a vanity metric, easily inflated by pricing or acquisition."

Evan stared at Lachlan. They hadn't acquired any company. They had not changed their pricing in a significant way. What was he talking about?

Lachlan turned to Jillian, Evan's CFO. "What is our

EBITDA per square foot of leased operating space?" he asked her, point blank. "Or better yet, what is the Adjusted Free Cash Flow per non-production asset? I'm lookin' for a different number," he said.

Jillian brushed her brown hair over her left ear as she looked at Evan. They had worked together to prepare for interruptions like this one.

Evan met Jillian's glance and silently mouthed the words, *"Hold. My. Beer."* Jillian nodded.

Evan held up his right hand, as if to stop traffic.

"Lachlan, I value your perspective, and you're absolutely right to focus on the bedrock of cost and asset efficiency. No one in this room, least of all myself, ever forgets that the fundamentals of business begin with Total Cost of Ownership. My dad always used to say, *'Sokutei wa taisetsu desu.'* That means: measurement matters. We are not moving the goalposts. We are talking about how we are going to score.

"In a company where the competitive edge is innovation, human capital is not fungible like a barrel of crude."

The oil business, Evan knew, was Lachlan's frame of reference. "Ours is not a standardized inventory item to be cut by a blunt instrument like 40%."

Gaining steam, Evan left no time for interruption or rebuttal. It was time to talk about the elephant in the room. And kill it, if he could. Because Evan had already measured twice before he spoke.

"To your point on TCO, we've used AI to eliminate the *cost* of those margins of inefficiency you spoke of. We've removed the low-value, repetitive tasks that were driving up our GSA and our IT spend. We are actively converting our cost of human capital from operational expenditure into strategic investment."

Evan shifted his tone. He addressed Lachlan like an old friend, even though he was anything but.

"Lachlan, you built your fortune on understanding yield. You understand better than anyone the difference between a high-cost, high-yield drill site, and a dozen low-yield pump-jacks. The initial Capex is higher on the drill site, but the return on invested capital is exponentially higher."

"RPE is our way of measuring the yield from our strategic human assets. We are moving from running low-yield wells to drilling a focused, high-output basin. We are not looking at the *number* of people; we are looking at the Return on the Value-Generating Capacity of those we keep." Lachlan sighed and looked down at his notes.

REGAINING CONTROL

"The next section of the presentation, which begins on Slide 8—the 'AI Integration Roadmap'—will give you the precise data, broken down by cost center, that quantifies this yield. It shows exactly how we have dramatically lowered the TCO of our *low-value* processes, while increasing the strategic investment in our *high-value* teams," Evan said.

"If you'll permit me here, I believe the data on this next slide will address your concerns about our focus on core profitability."

Financial Metric	Before Human Success (Projected Baseline)	After Human Success (12-Month Projection)	Impact
Revenue Per Full-Time Employee (RPE)	$275,000	$330,000	+20% Growth
Total Headcount (900 Baseline)	900 FTEs	830 FTEs	-7.8% (Strategic Right-Sizing)

Evan narrated slide 8: "This new structure is not about indiscriminate firing; it's about re-engineering value creation. We project a 20% increase in RPE over the next 12 months. This is achieved by having AI absorb the tasks that cost us 15-20% of our capacity, freeing up our talent to focus on high-value, revenue-generating activities."

Lachlan couldn't resist the chance to interrupt once again. "According to the data I have here," he said, flipping the pages in front of him until he landed on the one he wanted, "our competitors still have a higher revenue per employee, and significantly less headcount." He turned the portfolio around and lifted it towards Evan. In the small conference room, it was easy to see the name of the competitor.

"I thought you might bring up that company," Evan said, glancing at the outstretched portfolio. The Board Book contained the same metric that Lachlan had privately paid another research firm to identify. "But RPE, as you said, can be easily manipulated by aggressive pricing or even the outsourcing of core functions. The number tells us what a company sells. It doesn't say what the company keeps.

"I mentioned a nuanced approach in my opening remarks, and part of that intention is that we don't rely just on RPE,"

Evan said, advancing the slides to show what every board member, except Lachlan, was seeing on the Board Book pages in front of them. He nodded at Jillian. The CFO spoke. She said, "Instead of relying only on RPE, we focus on a metric that is far more relevant to profitability and shareholder value: Gross Margin Per Employee (GMPE)."

Evan continued, "If you look at the slide, you'll see that while their revenue is high, that competitor's high-volume, low-margin business model gives them a GMPE that is 15% lower than ours."

"Their high RPE is a revenue vanity metric built on a thin margin. Our 20% growth in RPE, however, has been deliberately coupled with a 10% growth in GMPE over the same period. We are not just selling more; we are selling more *profitably* because our AI integration is focusing human effort on high-margin customer solutions, while dramatically cutting cost-of-goods-sold and overhead."

Evan wasn't finished. "This brings us back to your point on the Total Cost of Ownership for our workforce. We are not chasing a headcount reduction that forces us into a low-margin race to the bottom. We are executing a long-term plan to ensure our employees are focused only on value-added tasks."

Lachlan nodded. All eyes were on him now.

"With your permission, Lachlan?" Evan continued. "Please turn your attention to the next slide everyone: The Voluntary Transition Program. I will detail exactly how we are achieving the necessary cost optimization while maintaining a positive market signal and upholding our commitment to our high-value employees."

Lachlan turned his portfolio around, sat back in his chair, and scratched his beard. His phone was nowhere in sight.

———

The Personalized Playbook

1. **The Vanity Metric:** The industry often focuses on Revenue Per Employee (RPE), which can be a vanity metric. What is your organization's current vanity metric (a number that looks good but hides a lack of profitability or efficiency)? What financially rigorous metric, like Gross Margin Per Employee (GMPE), should you be using instead to prove true, sustainable human-driven value?

2. **The Voluntary Solution:** Evan implemented a Voluntary Transition Program to address cost optimization. If your board or investor demanded a cost reduction, what **humane, value-preserving mechanism** (like a paid career transition program or a temporary sabbatical) could you propose to achieve savings without resorting to immediate, forced layoffs that would damage culture and talent retention?

3. **Context for Numbers:** Evan's presentation succeeded because he provided context for his numbers, linking augmentation to profitability. When you present metrics to your leadership, how clearly do you articulate the human story (the "why") behind the change in the number? Is your presentation about reporting or about persuading?

4. **The Pivot to Profit:** Evan focused human effort on high-margin customer solutions. Which three core activities are your most talented employees currently spending time on that are low-margin or non-essential? How could AI augmentation and a "Human Veto Rule" (Chapter 13) free them up to focus 80% of their time on the high-margin, strategic tasks that only they can execute?

5. **Earning the Right to Continue:** Evan had to earn his extended timeline and the right to propose a new plan. When you face internal resistance to a major cultural or strategic change, what non-negotiable, financially quantifiable outcome (a leading indicator) must you deliver in the first 90 days to earn the trust, attention, and necessary capital from your most skeptical stakeholders?

CHAPTER 17

THE CONFESSION

"The Human Success Playbook is fundamentally an EBITDA-optimization strategy," Evan shared. "We are not moving the goal posts here; we are showing that costs are relative to the value they create. After the break, the executive team will come up front here, and we'll address any specific questions. So, break time: 15 minutes everybody."

At the back of the room, several board members congratulated Evan on the numbers. The progress had been swift, with the Human Success playbook, and the entire board took notice. Almost.

Lachlan touched Evan's arm. "A word?" he asked. Evan excused himself from the dialogue with other board members, picked up his coffee, and walked out onto the patio with Lachlan. The curtains had been pulled back, and the floor to ceiling windows opened completely, creating a kind of indoor-outdoor room. The effect was classic Southern California. The two men walked across the Mexican tiles, past the open windows, and onto a veranda. Together, they looked out across the Los Angeles River at the Lion's Lighthouse, and the Queen Mary II.

"Ah, Glasgow," Lachlan said to no one in particular.

"I thought you were from Aberdeen?" Evan answered.

"Not me. The Queen Mary." Lachlan said. "Built in Glasgow. Well, Clydebank, actually."

He turned to face Evan.

"Nice presentation," he grunted. "But you're leaving money on the table. You're going to have to cut more headcount. And when you do, your efficiency is going to be even better." He looked out at the gigantic floating hotel with Scottish origins. He narrowed his eyes as he faced the CEO. "You know there are more cuts ahead."

"Maybe you're right, Lachlan," Evan said, taking a step to his right and setting his coffee cup on a nearby cocktail table. "Maybe we will get there." His tone was easy and loose, but he chose his words carefully.

"But we can't *start* there. I'll do whatever is necessary to grow this company. I'm not going to cling to headcount if it's not warranted. But we're moving forward with Human Success. That's how we are approaching our leadership. Our teams. Our collaboration with AI. No matter how many employees we have, that's not going to change."

Evan took a breath. He was tired of the pushback, the sharpshooting, the bullshit. It was time to find out what was really going on here. If Lachlan was going to cling to a mass firing, perhaps he should start at the top.

"Human Success is our approach. That approach is not going to change while I'm CEO," he finished.

Lachlan stared in silence, looking across the water. It was his move.

"I've got a confession to make," Lachlan said, facing the same direction as Evan and peering at the lighthouse. Boats passed leisurely in the waterway in front of the two men. The bright sunshine was offset by a perfect Southern California breeze. "I didn't think you could do it," he said.

"You're not done, I know," Lachlan continued. "But you are making progress. The numbers look good. I still believe that they can be better, but…they look good. This initiative around Human Success seems to be the right approach, and it is driving your AI integration in a way that is…thoughtful. Not disruptive. The numbers support the changes, and the employees are engaged as a result."

The remark surprised Evan, and he turned to look at Lachlan.

He wanted to say "Thanks", but that felt premature. Instead, he just nodded.

Lachlan edged closer to the CEO. The tone of his voice shifted.

"This is your show, Evan," he said. The men were face to face now, standing on the stone patio outside the conference room.

"Remember," Lachlan continued, "customers don't come

first. Employees come first. If ya invest in your employees, they will take care o' your customers."

Evan blinked. "That doesn't sound like you."

"Well, I can unnerstand why," Lachlan replied, with a shrug. "Richard Branson said it first. An' better'n me. But it's true. Look, I always knew ya would find the right way forward. I was counting on it."

Lachlan watched a jet ski zigzag across the harbor. He turned back to Evan. "I measured twice and I cut once." He smiled. Evan smiled back.

"You are steerin' this ship. Keep goin'. Keep producin'. Stay focused on the Rule of 40 and EBITDA. Look for ways to maximize this company - and maximize Human Success," Lachlan said. He checked his watch and looked back over his shoulder into the conference room. "Maybe this investment in people is tha' right thing to do. It's not all about AI. Growth and profit are tha' result of investin' in your people.

"You have my full support, Evan," Lachlan finished. "Don't screw this up." He turned and walked back into the conference room, leaving Evan alone and staring at the sea.

Evan closed his eyes and took a deep breath. He felt the warm California sunshine on his face. As he inhaled, he smelled the saltwater and the scent of the gardenias blooming nearby. He exhaled, slowly, and opened his eyes. Somehow the world looked different. Even though it was exactly the same.

He grabbed his coffee cup and returned to the meeting.

———

The Personalized Playbook

1. **Identifying Your Catalyst:** Evan initially saw Lachlan as his antagonist but later recognized him as the catalyst who convinced him to produce a superior strategy. Who is the most difficult or demanding critic (investor, board member, competitor, internal stakeholder) in your professional life? How might you reframe their pressure as a gift that is forcing you to perform your best work?

2. **The Pressure Test:** Lachlan's demands forced Evan to create a strategy that was financially and culturally bulletproof. If you knew your proposal would be reviewed by your most ruthless critic, how would you change your current AI or human capital strategy to make it financially indisputable and immune to emotional arguments?

3. **The Flip in Philosophy:** Lachlan's "confession" validates the mantra: "Employees come first." How must your company structurally change (e.g., compensation, budget allocation, decision-making authority) to prove to your workforce that this philosophy is not just a poster on the wall, but a governing principle for resource investment?

4. **Beyond the Binary:** Evan successfully forced Lachlan to accept both: "Maximize this company - and maximize Human Success." Name one area in your business where you currently believe you must choose between a financial outcome and a human outcome (e.g., cutting training vs. hitting profit targets). What structural element from the Playbook

(e.g., GMPE, Trifecta, Productivity Share) can you apply to prove that you can achieve both simultaneously?

5. **The End of Fear:** Evan's internal world is transformed even though his external world remains the same. What is the single biggest fear or uncertainty you have been carrying about the future of AI and your organization? What specific piece of governance or data (a "Human Veto Rule," a "GMPE" calculation) could you put in place this week to structurally neutralize that fear, allowing you to focus on strategy?

R asmus was first.

He was the first one to remove the headset, unlatching the apparatus at the back and slowly lifting the half-helmet mechanism off his head and face. He peeled off the medical-grade breathable silicone that covered his nose and mouth and set everything on the table in front of him. He turned and watched as Robin and Nina did the same.

The three executives squinted and tried to adjust to the lights. They were seated in the left front row of the all-white training room. They sat in ultra-comfortable white padded gaming chairs at a long thin table. The table was one of six large conference tables inside the room, organized in rows with an aisle in the center, classroom style. At the other tables, 18 other executives from various companies were pulling off their headsets - seated in the exact same kinds of chairs and configuration.

Robin was looking at the back of the room when an Asian man in a white lab coat began to speak. He turned to the sound of the voice, which was oddly familiar. The three executives instantly recognized him.

The man was Evan Sato.

But not exactly. This Evan Sato was… different.

He was older. Thinner. His voice was calmer, more grounded and certain. The uneasy Evan was nowhere to be found.

His hair was pulled back in a ponytail, but long strands of gray and black hair had escaped, surrounding his face and emphasizing the tiny clear glasses that rested on his nose. He wore a white lab coat with an unusual Nehru collar. The coat was more substantial than just a regular old lab coat - the fit was tailored and looked custom made. Two other technicians, wearing custom fitted lab coats in the same style, flanked him on either side at the front of the room. From his seat in the front row, next to Nina and Rasmus, Robin could easily read the name embroidered above the pen pocket: Dr. Evan Sato.

"What you have just experienced represents our latest advancements in experiential modeling for organizations, building on the work of Dr. Louis Rosenberg," Sato said. "Our patented Climate Parka solution is what allowed you to feel the breeze in New York and the rain in London."

Nina looked down at the white puffy coat she was wearing, just like everyone else at all the tables. Although the fluffy jacket looked and felt like something you would ski in, she was completely comfortable inside the climate-controlled suit. A mesh umbilical cord, no thicker than a garden hose, connected the left sleeve of every parka to a central hub in the middle of every table. She noticed that next to the central hub on every table was a raised silver logo with the company name: Veridian.

"Did you smell the gardenias in Long Beach? That's also part of the immersive sensory experience we've created for you. And for your customers," Dr. Sato said. Robin looked down at the table and picked up the silicone mesh face mask he had just removed. The mask was connected to the right shoulder of his parka via a clear flexible tube that was thinner than a pencil. The mass of tubes and interconnected technology was coming into focus as Dr. Sato spoke. The parkas and face masks had created the climate changes and sensory experience for all the participants.

Two technicians in custom lab coats wheeled in an 85" television screen with a video presentation already loaded and running. The Veridian logo spun and flew across a universe of stars as it transformed and disappeared into the streets of New York, London and Long Beach - recreating a flying drone-style video of the locations from the simulation. The arrival of the television presentation was seamless, as Dr. Evan Sato stepped forward and the technicians placed the television behind him on the left.

"Now, with Veridian, you can run comprehensive and immersive simulations of AI-human interaction, just as you have experienced today." The words "Immersive" and "Interactive" spun and danced on the screen behind him, as a video showed a server room bustling with technicians revealing the inner workings of the system. Dr. Sato continued, gesturing with his right hand to the front row of seats. "I want to thank you, Rasmus, Nina and Robin, for volunteering to be our beta testers today." Rasmus nodded, and Robin smiled. Nina, on the other hand, wasn't sure how she felt about being a guinea pig. She was still processing what had just happened.

The slide changed to reveal a single word at the top of the screen: PRICING MODEL.

"Your customers can customize the experience completely. Just as you could have ventured into other places, such as visiting the company headquarters or exploring the streets of New York, the solution is immersive and goes where you want to take it. If you had wanted to speak with other board members, or interact with other people within the various scenarios, or take The Tube to Kings Cross, you could. You would find that there are no NPC's here," he said, referencing the non-player characters that are often silent cannon fodder in video games. "Our simulation is complete, designed for your exploration, and that's what you receive for your investment. More importantly, you can resell the solution to your clients within your ecosystem, adding even greater value for your customers who are considering what human-AI interaction might look like. You are investing in a comprehensive digital world."

When Rasmus saw the price tag on the screen, he sat back in his chair. He wanted to keep a poker face, but he couldn't stop the frown. A murmur of gasps and whispers passed through the crowd behind him - the sound of sticker shock. Robin looked over at Nina, as he raised his eyebrows. Nina just stared, unblinking, back at Robin.

After several more slides and videos, the presentation concluded. Rasmus stood up and approached the front of the room. "There's a lot for us to consider," he said. "Is there a conference room or someplace where I can go with my team to discuss what we've just experienced?"

At Sato's request, a technician showed the three executives to a nearby meeting room so that they could speak in private.

Inside the windowless white conference room, there was a round table and four white chairs.

Robin couldn't sit down.

"That was awesome!" he said.

"I know," Nina agreed. Then, thinking better of her last remark, she blurted out, "What the hell was that?"

"It was amazing," Rasmus stated. "Super valuable. We've just seen the GPS that every company needs for guidance and governance in the age of AI. But there's no way in hell we could afford to invest in this technology. The price tag is outrageous. And what are we supposed to do, shift our business model into reselling experiences to our customers?"

"But think of the marketing implications," Robin said. "Talk about 'leading edge'!"

He sat down on Rasmus' left and gestured with both hands as he spoke. "We could show companies, using their own data, their own people and experiences, how to implement Human Success. Or rather, they could show themselves. We could create events for them in real-time, with a boots-on-the-ground experience of how to transform their culture and their leadership," he said, his voice rising with enthusiasm.

"It could be a great way to connect human resources to human success," Nina said. "Through personal application, I mean."

"Look, I hear you," Rasmus replied. "The experience was amazing."

Robin spoke up in agreement. "Yeah," he said. "It felt so real! Think of what this could mean to other companies that are going through a transformation around AI, and wondering how to keep people at the center of the transformation…It would be epic!" he concluded.

Robin's enthusiasm was contagious. But Rasmus had to be realistic.

"Look at the numbers. There's no way we can make this investment. No way to acquire this technology or even… I don't know… lease it? It's not companies that are going to buy in, it's *countries*. Governments. Wow. These numbers don't feel like something that even Microsoft could touch. And it would take years of business development for us to find early adopter customers. Our shareholders would never go for it."

He stopped to collect his thoughts and change direction. "Look, we all know that Human Success works. We know it can be applied to all kinds of businesses. Just like what we saw, we are figuring out how to work with AI in real time, right now. And I know there are thousands - no, more likely hundreds of thousands - of companies that are trying to get this right. Simulation and modeling make a ton of sense! But creating some kind of advanced immersive simulation just isn't feasible. I'd love to give every company their own Human Success experience - their own GPS for the age of AI. But…I'm sorry. There's nothing we can do."

Nina looked over at Rasmus.

"Actually, I just realized something," she said. "There is something we can do."

Rasmus raised his eyebrows. "I'm listening," he said. Robin leaned in and put his elbows on the table.

Nina looked back and forth at her colleagues. "The price tag is out of reach," she said. "But it doesn't have to be. We all need a GPS for the age of AI. The wisdom must be shared."

Robin couldn't wait. "So…How do we do that?"

"Well, you know," Nina said, "we could write a book."

+++ ***THE END*** +++

ACKNOWLEDGMENTS

This book would not have been possible without the vision and commitment of my co-authors: Rasmus Holst, Robin Daniels and Nina Carøe. Your insights into the human operating system informed the story and the creative process. Thank you for all of your contributions and conversations. Robin, seeing you live onstage was truly one of the highlights of our work together. Your trust in me has been the foundation of this book, and I am grateful for the opportunity to bring your ideas to life.

AI can never replace human creativity, and the creative process - but it can augment it. More importantly, I am deeply grateful for the humans who helped me in my journey to create this story - my 10th book since the pandemic. Thank you to Jules Swales, author of *Lies, Lies and More Lies*: your book helped me to find the truth. Thank you to Lorna Davis, for sharing your insights into the mind of a CEO - this book would not have been possible without your input. To Louis Rosenberg, Kristi Palma, Karen Mangia and Ken Orman: your brilliance helped shape this book, and I am deeply grateful

To my wife, Lisa-Gabrielle, and my daughters Ruby and Noli: your support throughout this project means the world to me. It all begins with you. I love you all very much.

Finally, to you (the reader): you're the reason we wrote this book. Thank you for diving in, and taking the journey. May it help you to continue your own, with greater ease and understanding of the brave new world that surrounds us.

Chris Westfall

ABOUT THE AUTHORS

RASMUS HOLST

Rasmus is the CEO of **Zensai**, where he has successfully transformed the organization from a specialized software provider into a global leader in the "Human Success" category. Under his leadership, Zensai (formerly LMS365) has scaled from $0 to over $30 million in ARR, earning prestigious accolades such as the  Red Dot Award for brand identity and recognition as a Great Place to Work. Rasmus is a vocal advocate for the evolution of workplace culture, arguing that traditional Human Resources must be rewritten for the AI era. His thought leadership centers on the idea of AI as a teammate—a tool that identifies learning gaps and removes obstacles to amplify, rather than replace, human potential. By integrating learning, engagement, and performance into a single real-time "Human Success Score," Rasmus champions a data-driven approach that prioritizes the growth and well-being of the individual as the primary engine for business results.

ROBIN DANIELS

Robin Daniels serves as a strategic leader at Zensai, bringing a world-class track record of scaling high-growth tech giants. With executive experience at Salesforce, LinkedIn, Box, and WeWork, Robin has been instrumental in 2.5 IPOs and has held three CMO roles throughout his distinguished career. He is known for his ability to blend Scandinavian resilience with a global mindset to build authentic, highly differentiated brands. His website: https://www.robindaniels.co/

Robin's philosophy focuses on "epic results" achieved through the empowerment of people. In the age of AI, he views technology as a bridge to deeper workplace connection and accountability. He emphasizes that AI should be used to align personal career development with organizational goals, ensuring that every employee has the tools and visibility needed to do the best work of their lives.

NINA CARØE

Nina Carøe is the world's first Chief Human Success Officer, a role she pioneered to signal a fundamental shift in how organizations value their people. Nina joined Zensai following the acquisition of Valuebeat, a company she co-founded to help businesses measure cultural impact through real-time human data. Her previous leadership roles at hyper-growth companies like Dixa and Unity Technologies have made her a sought-after expert in scaling international teams.

Nina's work is dedicated to putting people at the center of every process. She leverages AI and machine learning not as cold metrics, but as strategic tools for measurement and impact—enabling leaders to move beyond "lagging" HR metrics to proactive, human-centric insights. Her mission is to ensure that "Human Success" is not just a department, but a foundational strategy that turns employee fulfillment into a competitive advantage.

CHRIS WESTFALL

Chris Westfall is the author of *The NEW Elevator Pitch* (international best-seller), *BulletProof Branding, Leadership Language* and *Easier* (Wiley). As a ghostwriter, he has written 10 books - including a *Wall Street Journal* best-seller for a senior executive at Salesforce. A contributor to *Forbes* for seven years, his writing has appeared in *Entrepreneur, US NEWS and WORLD REPORT, SUCCESS, Fast Company* and many other print outlets. He regularly appears in national and international media, and provides leadership communication consulting to a wide array of clients.

Find out more: westfallonline.com or on social media:

instagram.com / westfallonline
facebook.com / therealchriswestfall
linkedin.com / in / westfallonline
youtube.com / @westfallonline

www.ingramcontent.com/pod-product-compliance
Lightning Source LLC
Chambersburg PA
CBHW040755120726
48005CB00012B/1189